TS
W9-ADZ-428
CAREER

TECHCAREERS:

Welding Technology

Joseph Abbott &
Karen Mitchell Smith

WITHDRAWN
KVCC KALAMAZOO VALLEY COMMUNITY COLLEGE LIBRARY

© 2011 TSTC Publishing

ISBN 978-1-934302-33-0 (softback)
ISBN 978-1-936603-05-3 (ebook)

All rights reserved, including the right to reproduce this book or any portion thereof in any form. Requests for such permissions should be addressed to:

TSTC Publishing
Texas State Technical College Waco
3801 Campus Drive
Waco, Texas 76705

http://publishing.tstc.edu/

Publisher: Mark Long
Art director: Stacie Buterbaugh
Editors: Mick Spillane and Ana Wraight
Marketing: Sheila Boggess
Sales: Wes Lowe
Indexing: Michelle Graye (indexing@yahoo.com)
Printing production: Thomson-Shore, Inc.

Manufactured in the United States of America

First edition

Publisher's Cataloging-in-Publication
(Provided by Quality Books, Inc.)

Abbott, Joseph, 1967-
Welding technology / Joseph Abbott & Karen Mitchell Smith. -- 1st ed.
p. cm. -- (TechCareers)
Includes index.
ISBN-13: 978-1-934302-33-0
ISBN-10: 1-934302-33-3

1. Welding--Vocational guidance. I. Smith, Karen Mitchell. II. Title. III. Series: TechCareers.

TS227.7.A23 2010 671.5'2'023
QBI10-600064

Table of Contents

Commonly Used Abbreviations

ANSI	American National Standards Institute
AP	Advanced Placement
ASA	American Standards Association (currently ANSI)
ASME	American Society of Mechanical Engineers
AWS	American Welding Society
CAD	Computer-Aided Design
CAWI	Certified Associate Welding Inspector
CDC	Centers for Disease Control and Prevention
CNC	Computer Numerical Control
CRAW	Certified Robotic Arc Welding
CRI	Certified Radiographic Interpreter
CSWI	Certified Senior Welding Inspector
CW	Certified Welder
CWE	Certified Welding Educator
CWEng	Certified Welding Engineer
CWF	Certified Welding Fabricator
CWI	Certified Welding Inspector
CWS	Certified Welding Supervisor
EPA	Environmental Protection Agency
GMAW	Gas Metal Arc Welding, including MIG and MAG welding
GTAW	Gas Tungsten Arc Welding, also called Tungsten Inert Gas (TIG) welding

HVAC	Heating, Ventilating, and Air Conditioning
MAG	Metal Active Gas, a form of GMAW
MIG	Metal Inert Gas Welding, a form of GMAW
MMA	Manual Metal Arc Welding, also called SMAW or stick welding
NDSCS	North Dakota State College of Science
NDT	Non-Destructive Testing
NIOSH	National Institute for Occupational Safety and Health
NRC	Nuclear Regulatory Commission
NTMA	National Tooling and Machining Association
OOH	Occupational Outlook Handbook
OSHA	Occupational Health and Safety Administration
OSU	The Ohio State University
QC	Quality Control
SCWI	Senior Certified Welding Inspector
SMAW	Shielded Metal Arc Welding, also called MMA welding or stick welding
STA	Safety Task Assignment
TIG	Tungsten Inert Gas Welding, also called GTAW
TSTC	Texas State Technical College
UV	Ultraviolet
Weld-Ed	The National Center for Welding and Training

CHAPTER ONE

Welding Careers

There are times when glue, tape, or a fastener – a pin, a nail, a screw, a rivet – just will not do. A bond has to be watertight or airtight; it has to hold under enormous stresses – the weight of a skyscraper, the pressure of a boiler. The materials involved have to be fused together into a continuous whole.

They have to be welded.

A *lot* of things have to be welded, from the smallest piece of jewelry to the largest structure. The work requires precision, rigor, thorough knowledge of the materials involved, and a lot of skill and practice because lives – sometimes thousands of lives – may depend on the quality of the weld.

It is precisely that need for skill and experience that is putting some industries, highly dependent on welders, in a bind. Most experienced welders come from the Baby Boomer generation, and they are starting to retire – with far fewer younger welders entering the profession to replace them, due in part to the disappearance of high school vocational education classes with welding shops. The U.S. Department of Labor forecasts as many as 400,000 (or more) welding positions will be vacant by 2014, but there will not be enough qualified welders to fill them. Already, the National

Tooling and Machining Association (NTMA) reports that forty percent of member companies are turning away customers because there are not enough welders to do the work.

"We need welders like a starving person needs food," Hal Connor, human resources manager for JW William, Inc., declared to the *Wall Street Journal*. So desperate is the maker of dehydration and compression machinery for welders, it began offering $1-an-hour bonuses to welders simply for showing up on time for work.

Bechtel Corporation's Bob Alexander adds, "The problem is the average age of a welder in the craft today is fifty-two to fifty-five years old, which is not when you peak. Welding is eyesight, reflexes. There is a whole generation of welders missing, and we're trying to fill that gap. Most of our new hires are in their low-twenties to mid-thirties."

One reason for the gap is the image of welding as grimy work – an image that is far behind the times. "It's not the blacksmith, dungeon type of work people have always assumed it to have been," says Dave Cotner, head of the welding degree program at Pennsylvania College of Technology. "We are seeing more and more technology coming into play, so it doesn't have the stigma of welders being high school dropouts that it once had."

Cotner goes on to explain that welders in today's marketplace have to be technologically savvy, as well as manually skilled in their craft. "The career is undergoing an image face-lift. That's a solid, positive thing. I tell students and parents: 'The shirt you are wearing is not welded, but it is tied within a few short steps to the machines that made it, the trucks that carried them, and both of those things were single welded.' It's a 'six-degrees-of-separation' thing. If we made all the welded things go away, and the results of those welded things, we'd be standing in an empty field, naked. The craft is getting recognized more as a science and the art it can be, rather than the black and white pictures of welders with blackened faces, wearing tight goggles, looking like they came out of the worst place in the world. With safety standards and ventilation systems, welding can be a very glamorous career."

Welding is a profession that can take you around the

world and even into the depths of the ocean, working with materials from plastics, steel, and the most exotic of metals. The techniques can be centuries old or cutting edge; a welder may even be called upon to invent new techniques.

While welding is traditionally a male field, the number of female welders is growing and expected to grow even more in the future as the desperate need for skilled welders trumps sexism. "It's a rough trade for women, but they tend to be extremely good welders. Women are particular about how things look," says Dan Barrow, Zachry Construction Corporation craft training supervisor. "A weld that looks good is good, normally. And one of the best pipe-fitter foremen I ever saw was a woman. She could keep two welders busy."

For a new worker with the right attitude and the right education the welding profession is a seller's market. The opportunity for advancement is nearly endless.

"I have one student working for a boilermaker," Cotner says. "He gets a bonus of $75 a day just for showing up before 7:00 a.m. The need is out there, and we don't see this stopping anytime soon."

Profile: *Jason Praster, Bechtel Corporation*

When Jason Praster came to work for Bechtel Corporation in 2006, he came with both an associate degree and a bachelor's degree in welding engineering from Pennsylvania College of Technology. Praster works within the power plant division of Bechtel, and the travel involved with the job is just one of the things he enjoys about working for this company. Currently, he is involved in an air quality control unit project in Stratton, Ohio, tying all units into one in order to eliminate smoke contaminants and bring the units up to Environmental Protection Agency (EPA) guidelines. However, in his two years with the company, he already has lived and worked in Nebraska, Texas, Arizona, as well as Ohio. With so much traveling, Praster, who is engaged to be married, believes he soon will enjoy another of Bechtel's benefits. "Bechtel pays for the cost of you to fly home or for your wife to fly out [to your location] once a month," says Praster. "This is a family-oriented business. It's owned by a family, and they look out for the family."

Praster currently works as a welding inspector, doing quality inspections with both visual and liquid penetrant testing of certain welds. "I also support construction as far as welding needs, making sure they meet codes and Bechtel procedures, and I support the rod room," Praster says. "Another of my jobs is to monitor a welding test lab, making sure we have quality welders in the field." Praster also carries responsibility for managing welding qualification and stamps, keeping them current and updated. "A welder has to weld every six months to stay current," he explains, "and if they're in Quality Control (QC) or some other area, they may not be doing as much hands-on work."

Although Praster's job description is considered non-manual, he is very much a hands-on person. He studied welding engineering at Penn College, where welding engineers spend their first two years learning manual welding skills. "When I graduated, I knew how to walk the walk; I knew how to strike an arc, and I knew how to talk to struggling welders," says Praster. "That has helped me succeed, and put me where I am today. Welding engineers have welder's hands with an engineer's mind. I keep that in the back of my head every day at work. When I'm out in the field, I think about that. I think, 'This guy is struggling and not picking this up. I learned this, so I teach him this way.' I get satisfaction from being able to help people out in the field. I get a personal satisfaction in explaining what they are doing wrong and showing how to do it right."

From the one-on-one development of welders to the massive projects he works on, Praster enjoys his job and sees his work as an important contribution, not only to Bechtel, but also to the power industry at large. "I've completed a couple of record-setting projects," Praster says. "At Fort Calhoun, a nuclear plant in Omaha, we did the largest scope of work ever done for a steam generator replacement project. We replaced the reactor head vessel, the pressurizer, and two steam generators in what they called the `Big Outage.' It was the largest scope of work ever completed and took us about sixty-seven to seventy days. We worked two shifts a day, working six days a week. It brought Bechtel a lot of recognition."

Another exciting project Praster worked on was at the Comanche Peak Nuclear Power Plant in Somervell County, Texas, where Bechtel replaced four steam generators. "We broke the world record for completing that job," says Praster. "We did it in fifty-five days."

He highly recommends a career in welding to any young person looking for adventure, travel, job stability, and excellent pay; however, "Make sure this is something you'd definitely want to do," warns Praster, "because there's a lot of travel involved. Does that fit your lifestyle? I had to make sure I really wanted to travel; I knew I'd be away from home a lot." Praster also tells prospective welders, "Be excited about the career. Study hard. Have fun with it. Welding is something that can be really fun."

More than just fun, welding is a lucrative job that offers endless career possibilities, says Praster. "There's always some new process; they're always inventing something new, more efficient. With such a shortage of welders, I see it [the career] going toward robotics and orbital welding – primarily because we don't have enough hands-on welders. You don't have to have a good hand at welding to do orbital. There are just unlimited possibilities. You can work at manufacturing, construction, NASA, etc. If you think about it, the whole world is based around welding. The structures, your vehicles, computer chips. Even plastics are welded now. There are lots of open positions."

According to Praster, the most important advice he has for a potential welder is: "Be prepared to learn something new every day. It's such a broad field – you can never ever learn it all."

Overview

Welding is a means of joining two materials into a continuous whole – usually, though not always, metals and typically, though not always, by melting them together. (Welding's cousins, soldering and brazing, join materials using a filler that is heated enough to melt the filler but not the materials being joined.)

Welding has been around for thousands of years. For most of that time, however, it took a form not many people today would recognize as welding: the blacksmith heating metals in his forge and hammering them together, then repeating the cycle. It was only in the last century or so that modern techniques began to appear, sparked by the start of the Machine Age in the late nineteenth century.

The electric arc – the superheated plasma produced when electric current flows through air between two electrodes – had been discovered early in the nineteenth century by the Russian scientist Vasily Petrov, who proposed its use in welding. However, it was not until eighty years later that another Russian, Nikolai Bernardos, invented the first arc welder, using carbon electrodes. Soon after, Russian inventor Nikolai Slavyanov and American inventor C.L. Coffin developed welding machines with metal electrodes. Around the same time, American inventor Elihu Thomson developed resistance welding, in which electric current flows directly through the metals to be joined, rather than forming an arc.

Oxy-fuel welding, also developed at the turn of the nineteenth century, was the chief competitor to electric welding, using a torch that combined oxygen with a fuel gas such as acetylene (discovered in 1836 by the British scientist Edmund Davy and named a quarter-century later by French chemist Marcellin Berthelot). Thermite welding, in which the heat is produced by a chemical reaction, also appeared at about this time.

As welding techniques developed during the twentieth century, one of the bigger challenges was shielding – finding a way to keep oxygen and nitrogen in the air from causing impurities in the weld. The first solution was to coat the electrode in a solid material – a flux – that vaporized to create a gaseous shield at the weld point, creating a layer of molten slag over the weld. (The electrode itself melts as part of this process, forming filler material for the weld.) This is called shielded metal arc welding (SMAW) or manual metal arc (MMA) welding. (A more colloquial name is stick welding.)

Another solution was a nozzle accompanying the arc-welding tip that surrounded the weld area with a shielding gas. Gas tungsten arc welding (GTAW) – commonly called tungsten inert gas welding (TIG) – uses an inert gas such as helium or argon to shield a tungsten electrode (since the tungsten does not melt, this usually requires the addition of a metal that does melt to provide filler). Gas metal arc welding (GMAW) returns to a consumable electrode, wire continuously fed to the tip, which provides the filler as well as the arc; this method is called metal inert gas (MIG) or metal active gas (MAG) welding, depending on what kind of gas is used for shielding.

From these methods springs the traditional image of the welder in a helmet and a safety suit wielding a welding rod. These days, however, a welder is almost as apt to be sitting at a computer console operating a remote device (especially in hazardous environments such as nuclear reactors). The welding itself may be done through newer technologies such as lasers, electron beams, ultrasound, or extremely high pressure (called explosion welding). Which of all these methods is chosen depends in part on the materials being joined and how the item being welded is going to be used.

Profile: *Andy Wolfskill, Acute Technological Services*

Andy Wolfskill's job has taken him around the world, but it all began back in high school with a welding machine in his best friend's garage. Wolfskill elaborates on his early beginnings as a welder: "We built go-karts and mini-bikes. He was good, and I got interested and did a little bit, too." By the time he was a senior, he was part of a vocational-technical program where he snagged a job after lunch welding at a trailer factory, building cattle trailers. "I was the gate person. They had a jig where you put all these tubes in and made a gate. I really enjoyed that," he says. After high school, he enrolled at Texas State Technical College (TSTC). The path to his senior welding technician position at Acute Technological Services was filled with long hours and challenging projects. First, at Brown and Root, he spent ten hours a day under the hood working on various projects, including a nuclear power plant in South Texas, where he learned and honed the orbital process, welding the loop piping coming off the reactor.

The next stop on his career ladder was a stint at Welding Services based in Atlanta, where he did maintenance and repair on nuclear reactors. "We did outages all across the country," Wolfskill says."These were high radiation areas. We had to do remote welding using a TV screen. We would get there, set up the welding equipment, and do mock-up training in a safe zone. We had local people, called jumpers, whose job it was to set up the equipment in the reactor and then get out. They were allowed to be in the reactor itself for only five to ten minutes, so they spent a week practicing their timing. We welders helped them get everything ready. Then it was all high tech – welding from inside the trailer using joysticks and cameras. We kept radio communication with the jumpers, who made changes as needed."

From there, Wolfskill moved on to owning his own rig, working for himself. "I made a lot of money, but it was all up to me. The taxes, the record-keeping, everything. I worked long hours with no benefits."

Thirty-four years after it all started for him, his dues paid, he is reaping the rewards of being a senior welding technician at a major corporation, which he describes as the best place he's ever worked. Most of his efforts are now focused on research and development with the exotic metals for which Acute Technological Services is known. Oil companies bring new materials that have never been welded and new designs that have not been proven. They need someone to weld and test them, so Wolfskill is the go-to guy. "My job is to look at the problems and try to solve them. We do procedure development and analysis. There are lots of new, high corrosion materials coming out that oil companies want integrated, and they want to know the best welding procedures. Acute has a big engineering group that helps us with that." When someone who knows how to weld exotics is needed, Wolfskill is one of the first to be called. He travels six to eight months out of the year. From a shut-down offshore oil rig in Nigeria, waiting for Wolfskill's expertise to make the needed repairs that will bring it back online, to three weeks in Italy, he enjoys those non-monetary perks mentioned by his operations manager, David Pratt.

While traveling with the job brings excitement to Wolfskill's life, he finds his greatest job satisfaction in training young welders.

As both a Certified Welding Inspector (CWI) and Certified Welding Educator (CWE), he takes his role as mentor and role model to these young welders seriously. "We do a lot of small bore tubing – 3/8 to ½ inch tubing … it's very difficult. These guys are catching on and doing good work." The welders who do the best work get noticed and make more money, with plenty of opportunities for overtime at the typical rate of time-and-a-half. In fact, according to Wolfskill, a six-digit income is not out of the question. Wolfskill says, "It is possible to make $100,000 a year after 10 or 15 years. This is when you can be sent out on jobs and not have to be monitored real closely. Your rejection rate is very low. You do good work, and we know you're versatile. Maybe you do lots of international jobs. Once you get to where we can send you out, the sky is the limit."

For the person just getting a welding career started, Wolfskill, who hopes to become more involved in the training end of welding, offers this advice: "Always do your very best, so you can go as far as you want. Attitude is everything. Keep a humble attitude and try to learn as much as you can. If an employer sees that in a person, he or she will get a lot more chances than the guy who thinks he can do anything." He also recommends that high school students planning to go into welding take as much drafting as they can at the high school level.

You might think a man who has welded everything from NASA space projects to offshore rigs in Nigeria to nuclear reactors may just want to relax and put his feet up when he's at home, but that's not the case. Wolfskill creates custom motorcycles as a hobby and says welding is more than a career for him – it's a passion. "I look at welding as sculpting. A weld is a sculpture you are making. You have to make it as clean and pure and even as you can. It's an art form. My mom still has the candlesticks I made her way back in shop class sitting on her dining room table."

Employment Outlook

Welders are needed everywhere in practically every industry. Welding skills are seldom so specialized as to only apply to one area. Even automation has a limited effect on the job

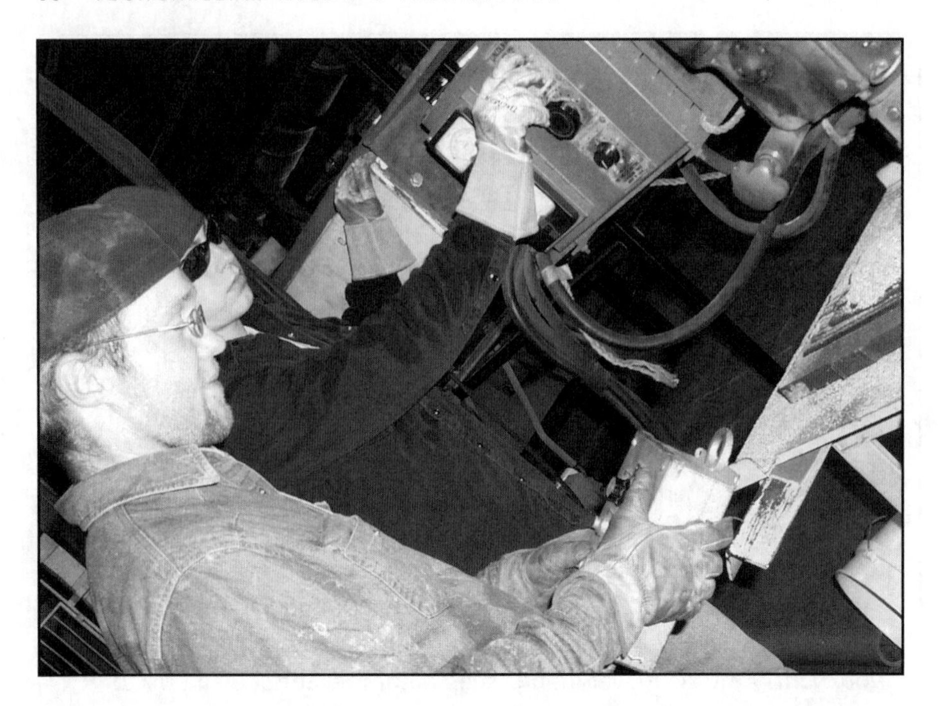

market – a weld made by a robot still has to be examined by a human eye for defects and machines adjusted accordingly.

The job market is wide open for those looking for a job. "We graduate about eighty to ninety welders a year," says retired TSTC Waco welding program director Frank Wilkins, "and we have one hundred percent of them placed in jobs, usually before graduation. Their job stability is excellent – companies hang onto their good welders and give them plenty of overtime and opportunity for advancement."

The U.S. Bureau of Labor Statistics' *2010-11 Occupational Outlook Handbook* (OOH) foresees little or no change in total employment numbers but says skilled welders should find work easily. "Job prospects for welders will vary with the welder's skill level," it reports. "Prospects should be good for welders trained in the latest technologies. Welding schools report that graduates have little difficulty finding work, and many welding employers report difficulty finding properly skilled welders. However, welders without up-to-date training may face competition for job openings. For all welders, prospects will be better for workers who are willing to relocate to different parts of the country."

According to the OOH, in 2008 there were about 412,300 jobs in welding, cutting, soldering, and brazing, and about 54,100 and more jobs running machines that did such work. About sixty-five percent of those jobs were in manufacturing, with concentrations in building construction and manufacturing of metal products, transportation equipment, machinery, and architectural and structural metals.

More than a dozen OOH job categories include welding, and a common thread runs through OOH discussions of their outlooks: Welding training and certification provide key advantages both in getting hired and in advancing.

While construction jobs in general are subject to economic whims, one area expected to see enormous growth in the near future is the nuclear industry, benefiting from the push for "greener," less polluting power generation. After a four decade lull during which few new plants were built, the Nuclear Regulatory Commission (NRC) is now considering applications for twenty-eight new plants at eighteen sites across the country.

Profile: *Bob Alexander, Bechtel Corporation*

Bob Alexander, manager of welder resources and training, has been privileged to work for Bechtel Corporation since 1971 when he started with the company as a welding engineer on a Colorado nuclear plant construction site. Over the years, including a nine-year period when he worked for a different company, his welding and management career has taken him to many locations within the United States and abroad, including a position as project field construction manager at a site in Saudi Arabia in 1997. He was also project manager on the Baghdad South power plant in Iraq in 2004. "That is what this job will do for you," Alexander says. "I've lived and worked in every state except Hawaii and Washington and all over the world. If you want to be a welder you better be willing to travel; they're not going to build a power plant in your backyard."

After completing the Baghdad project, Alexander intended to retire, but Bechtel had other plans for him. Due to the impending work crisis caused by the national welder shortage, the company asked him to get involved with managing, tracking, and recruiting welders on the corporate level.

Alexander's and Bechtel's efforts are paying off, and nationally, companies are responding to the welder shortage with higher paying jobs. "In the last year, wages have gone up about $10 an hour. For a long time you could make as much money managing at McDonald's, but now the pay is better," says Alexander. "At Bechtel, we are now getting more candidates than we can accept in our schools. Every student that has graduated from our school gets a job with us. We've been able to keep about eighty percent of our graduates still employed with us after six months. The people who go through our skills and successfully complete [the training] are finding work and sticking with it. This is good for the trade. People think, 'if other people are staying in, maybe I want to.'"

As a part of Bechtel's ongoing mission to help mediate the welder shortage, the corporation partners with several colleges in providing grants to help pay for welding consumables such as rods and gases. Alexander offers this advice to students in those programs, as well as to anyone else considering welding as a career, "Learn your trade, whatever it is. Be one of the best. Show integrity and commitment to doing a good job. Everything you do builds your reputation, no matter what kind of work. Have integrity, take pride in what you do, and you will develop a great future in the business."

Salary Ranges

As with nearly every job, welders' pay varies with factors such as location, experience, education, and the required level of skill.

Median wages for jobs in welding, cutting, soldering, and brazing, as of May 2008, according to the 2010-11 OOH, were $16.13 an hour, with the middle fifty percent earning between $13.20 and $19.61 an hour. The top 10 percent earned more than $24.38 an hour. Among those operating welding machines, the median wage was $15.20 an hour, with the middle 50 percent making between $12.62 an hour and $18.63 an hour, and the top 10 percent making more than $23.92 an hour.

Wages do not tell the whole story, however. "Although about 50 percent of welders, solderers, and brazers work a 40 hour week, overtime is common," notes the OOH, "and about 1 out of 5 welders work 50 hours per week or more."

Training and certification add significantly to pay. Dan Barrow, a craft training supervisor with San Antonio, Texas-based Zachry Construction Corporation, says a newly hired certified welder at Zachry starts with the journeyman's rate of $22 to $26 an hour, depending on location. "As the welders increase skill level and gain new processes, a pipe welder may get an extra $1 to $2 per hour. The X-ray welders may be at $2.50 above the journeyman rate. We also offer insurance and a 401(k)."

Bechtel's Bob Alexander says a top-of-the-line welder, doing pipe and alloys, can make $27 to $38 per hour at Bechtel. "Most of the jobs I know of are more than forty hours a week, so you get overtime," he says. "Most companies pay per diem for the travel. There is good pay in the business, and it is getting better all the time because of the shortage. There is a lot of opportunity out there."

At the highest end of pay are welding engineers. Materials engineers, which include welding engineers, had a median wage of $39.34 an hour in May 2008, according to the Bureau of Labor Statistics in the 2010-11 OOH; the top 10 percent made more than $59.84 an hour.

Job Titles

Job titles can depend on specialization or on level of responsibility. Different titles within the welding profession, besides the generic welder, include the following:

- **Boilermaker** – builds and maintains metal vessels and structures such as boilers, tanks, furnaces, smokestacks, and heat exchangers

- **Pipefitter** – installs and maintains pipe systems and pipe supports as well as related hydraulic and pneumatic equipment
- **Reinforcing and structural iron worker** – installs girders, columns, and other construction materials, including the metal reinforcement within concrete
- **Sheet metal worker** – makes, installs, and maintains items made from sheet metal, ranging from heating, ventilation, and air conditioning (HVAC) ducts to gutters and siding to outdoor signs, tailgates, and precision equipment
- **Underwater welder** – works on underwater projects such as offshore oil platforms, requiring experience in diving as well as in welding
- **Welding machine operator** – sets up and tends welding machines
- **Welding educator** – teaches welding techniques and principles
- **Welding sales** – sells welding equipment and supplies, often demonstrating their use

Three titles that can combine with any of the above deserve particular attention.

Welding Engineer

The job that requires the highest level of education within welding is the welding engineer, who must have at least a bachelor's degree. The welding engineer must be thoroughly versed in metallurgy and engineering principles. In addition, they are responsible for developing, evaluating, and improving welding techniques, procedures, and equipment, especially when it comes to new alloys or unusual fabrication methods or materials.

The welding engineer is as much scientist as welder – keeping abreast of new developments in the field as well as developing hypotheses, conducting experiments, and publishing the results.

"Welding engineers spend most of their time behind a computer screen, writing reports, doing calculations," says Dr. Charles Albright, an associate professor in welding engineering at The Ohio State University. "When they go out into the job site or into the lab, most of the time they are not very involved with manual skills welding." Albright adds many times engineers are working with electron beams, lasers, or resistance welding. "There is a whole world of welding that does not require manual welding. Manual skills are vitally important, but it takes a whole different set of skills to operate a laser welding system."

Welding engineers must often be instructors and teach others the new techniques and procedures. They also will be in supervisory positions, including certification of subordinates' qualifications and inspection of their welds.

Welding Technician

The welding technician holds something of a medium-level job between entry-level welders and the welding engineers. Again, just as a researcher depends on the technicians who maintain and operate lab equipment, the welding engineer depends on welding technicians to conduct experiments and evaluate the data. Welding technicians also perform field testing and recommend materials and techniques to engineers. They, too, often will serve as welding inspectors.

Welding Inspector

In some ways the welding inspector, often a welding technician or engineer, is the integral link in the chain. In fact, the inspector's job is to seek out weak links and strengthen them – the editor to the welder's writer. This second set of eyeballs ensures that welds are consistent and meet contract specifications and legal requirements.

Many projects require – by law, contract, or both – that a certified welding inspector sign off on the work. This can mean both travel and profit for inspectors who contract out their services.

Doing the job right requires tenacity, strict attention to detail, and a strong sense of ethics. Never forget, especially with work such as structural welds or welding on high pressure vessels and pipes, lives are at stake. An incomplete job – or, worse, looking the other way under pressure from a client who does not want the expense of repair work – can kill. "Whether you function as welders or inspectors," says the welding-instructor hero of Lois McMaster Bujold's novel *Falling Free,* "the laws of physics are implacable lie detectors. You may fool men. You will never fool the metal."

Profile: *Russell Battles, Fluor Corporation*

When Russell Battles decided to become a welder, it came as no surprise to his family. His grandfather, his dad, and his uncles were all welders. In high school, Battles was taking welding as part of his agricultural classes. After high school, he started out at a local community college, but soon realized academia was not what he was looking for. He enrolled at TSTC, and continued his welding training there.

Eighteen years later, Battles is a welding inspection manager for Fluor Corporation. He currently works in the company division contracted to Luminant, the electric generation arm of what used to be TXU Corp., for maintenance of all facilities. His responsibility on each site where he works is to set up and ensure the QC inspection program for all welding. "I also set up the boiler inspection programs and train the inspectors," Battles says. Additionally, he ensures that all the work his welders do meets code.

As with many welders and managers, the domestic and exotic travel involved with the job is one of Battles' favorite perks. "What I like about it is the opportunity to travel. You can see the country or see the world and get paid to do it," he says. "I left Texas in 1999, spent five years on the East Coast and could have gone overseas if I had wanted. Fluor just finished a five-year project in Kazakhstan that I could have been a part of."

Today, Battles sees his greatest challenge quite differently – specifically, a manpower challenge. "The hardest part is finding the skilled individuals to do the job. Technical trades have been pushed to the wayside, so there are not enough trained people," says Battles. "We are already seeing major welder shortages, and it affects us in being able to complete projects on schedule and under budget. We have to hire subcontractors that staff nothing but welders."

As welder recruitment efforts at Fluor ramp up, Battles offers the same advice to potential welders today that his own father gave him eighteen years ago when he told him he planned to follow the family tradition of professional welding. "The first thing my dad said when I told him what I was going to do was, 'if you learn a trade with your hands, nobody can ever take that away from you. You can go from there and complement it with academic education, but you'll always have a way to make a living.'"

Career Paths

As a welder gains skills, experience, and education, the options for the future open up substantially. "You can start as a welder, but there's no limit to where you go in the business," says Bob Alexander.

There is, of course, advancement within the job itself: Greater experience brings more complex or technical work (and the higher pay that goes with it). Promotions come easily for those who excel in their craft and work ethic, says David Pratt. "They are the first to be called on to become lead technicians," he says. "If we assign a crew of four or five or more, the ones who excel will be the lead representatives."

Welding is an art where there is always something to learn – and something to teach. Experienced welders can move to positions where they pass their knowledge along to younger colleagues, either within a company's training program or as a high school or college instructor.

Moving into welding research as a welding technician is another option, and with further college education, a technician can become an engineer. A welder also can go on to become a welding inspector, double-checking the work of other welders (as a formal job, a welder in a supervisory position will be doing this for subordinates), or put his or her skills to use in sales, working with welding equipment or supplies.

Finally, those who want to be their own bosses have plenty of opportunity within the welding profession – opening their own welding shops, hiring out for contract field work, becoming independent welding inspectors, or launching their own equipment or supply business.

Bob Alexander offers some sage advice for potential welders who plan to have skill training only. "If you have the skill for welding, that is a trade. But if you have the opportunity to take night courses for a degree, do that, too. As your physical skills diminish, be developing mental replacement skills that let you move into management with the same company you have been working for."

Profile: *Andy Thomas, Zachry Construction Corporation*

Andy Thomas has found Zachry Construction Corporation to provide everything it promises and more during the last seven years he has worked with the company. A native of a small town in Tennessee, Thomas grew up in an area where most men farmed, cut wood for the local paper mill, or went to welding school. He opted for welding at Athens Vocational School. After training as a steam pipe fitter and welder, he went to work for the Tennessee Valley Authority. Through the years, he worked on various projects for other companies, rising from foreman to general foreman, pipe superintendant, site manager, and then start-up engineer. Next came turnover coordinator and, now, field quality control manager at Zachry Construction Corporation. He also teaches welding three nights a week and recently had the privilege of seeing his own son pass his first welding test.

He offered the same advice to his own son as he does to any young person considering welding as a career. "Are you a watcher or a doer?" he asks. "If you're a watcher, there's no need to train for a career in welding because it's a competitive world. Some think they are good, some know they are good, and some show they are good every day in the way they dress, the way they go about doing their job. Can you handle making $100,000 a year and traveling coast to coast?" According to Thomas, if a person meets those criteria, then welding is an exciting, unlimited career.

During Thomas' twenty-eight years in the craft, he has seen most of the United States and has been involved in several projects that, in his mind, stick out as spectacular, even dangerous, events. For example, in 2000 he was on a project in South Texas where a fuel truck carrying 4,000 gallons of fuel caught fire as it filled storage tanks. The entire site burned, but with Zachry, as with all major companies, safety is a precision-drilled science, and quick reaction and thinking carried the day.

In a reclaimed mangrove swamp in Florida, Thomas helped build a major power plant, which he describes as "beautiful, rising out of the swamp as you look out across Biscayne Bay with the sun coming up."

The pride Thomas takes in his work is evident, yet he humbly says, "Learning to weld is like playing golf. Every pro you ever talk to will tell you they could do something to help their game. Every welder goes to work every day knowing there is one little something they can do to make their next weld better than the last one."

Thomas' day begins early, sometimes at 1:30 a.m. A quick shower, a kiss to his sleeping wife, and he is out the door by 2:30 a.m., headed to his office. Other days start a little slower. By 6:00 a.m., he is on site, meeting with his inspectors, and going over the Safety Task Assignment (STA) of the day – walking the site with inspectors, visiting with his welders – then back to the office by 8:00 a.m. for code document reviews. "At times, it seems I will never catch up," Thomas says, "and at other times, it is the most rewarding job in the world. Welding has not changed much over the years – we are just welding more exotic metals. You still work hard at a hot job, playing with fire, and hiding your face to make a living. It's just you and that little blue light."

Thomas' passion for his career is something he hopes to spread to the next generation as he teaches new welders. "I would advise young people to try it," he says. "If they like working with their hands, and getting personal satisfaction, I know that if they have done their best every time they finish a weld, it will become a passion."

Job Duties

For basic jobs, welders simply weld – pieces already have been planned and laid out for them. As jobs get beyond that basic level, welders start having to do the set-up work themselves.

That means being able to read blueprints and specifications and knowing how to get from "here" to the "finished product." You must know how to properly align and secure pieces, as well as know what the machines and metal can or cannot do, selecting procedures accordingly. You will need to know how to cut metal as well as join it. You also are expected to clean up afterward, grinding or cutting away rough parts.

Welding requires not just a steady hand, but a critical eye. You must know your work meets the standard – again, lives may well be depending on it. The simplest tests use measuring instruments and the Mark One Human Eyeball, but there are also means to extend its power: hydrostatic tests, X-rays, and ultrasound.

At higher levels – welding inspector, technician, or engineer – you will need to be familiar with the means of non-destructive testing (NDT) and know how to interpret the results. (Before this method, the only way to test a weld was to see what it took to break it.) There also will be paperwork; you will need to write up the results of your tests and keep records of them. Even more than the lower-level welders, you will need to be familiar with industry codes and legal requirements. Furthermore, as a welding technician or engineer, part of your job may be to test your employees as well as their welds, to make sure they meet certification requirements.

At the highest levels, welding engineers have to get creative, putting together knowledge of materials science, physics, chemistry, math, and electricity to meet the new needs of ever-changing technology.

Profile: *Jimmy Owens, Fluor Corporation*

Although Jimmy Owens has been with Fluor Corporation for only one year, he actually completed his welding education at TSTC in 1994, at which time he went to work for a government-contracted satellite antenna manufacturer in Wortham, Texas. "We did some big jobs there," says Owens. "I remember one about five years ago where the top brass from every military branch came to check things out. I had just gotten married and had to leave my honeymoon early to start that project. The company tested more than a hundred welders for the job and chose just me and five others. We welded on a piece of metal that averaged 1,200 degrees, and I sat against that thing for three months." Owens explains that the crew could only work for a maximum of an hour and a half before they were forced to take hydration and rest breaks. The antenna eventually went to a refurbished offshore oil rig. Owens says, "The satellite dish itself was a football field wide. Six of us did that thing. When it was all in place, the guy who engineered the design took us down to the coast to see the finished result. That is one of the most important jobs I have ever worked on."

At Fluor, Owens is part of the crews contracted to a Luminant power plant in East Texas. He had originally planned to leave the welding career for another direction, but the career kept calling him back. "It comes easy to me," Owens says, "and I care about my work. It does not matter whether I am welding at a power plant, on a satellite antenna, or on an old farmer's fifty-year-old plow, I make a good weld that is going to hold. My bosses at Fluor know they can walk away from me, and I am going to do the same quality of work as when they were standing there watching me. I have been doing this since 1993, and I have done hundreds of welds from stick to TIG to MIG. I do not weld trash. It's just clean-as-a-whistle welding."

The pride Owens takes in his work is what drives him to work eight hours a day for Fluor and then go home and work for himself. "After I get off my regular job, I weld fences, pens, and farm equipment. When I weld an old plow, it's going to look like it came out of a fabrication shop. I do not weld to make it pretty – I weld to make it right, but pretty comes as a part of that. My work does not break. I take my time and make it right, but I do not charge a lot. I have a lot of people stay after me to charge more, make more money, and travel. But I have two little girls I do not want to be away from, so I stay close to home," Owens explains. "I love to hunt, so I trade a lot of my welding work for hunting privileges, instead of charging more."

In the future, Owens would like to move into welding inspections or even possibly instruction. "Young kids come to Fluor to learn and test. I would not mind being an instructor. I have a lot of knowledge I could pass on," he says. "I took what they taught me at school, brought it to the job, and then learned a lot of tricks from the experienced guys on the job. I would like to do that for the kids who come in here."

Safety

Bob Alexander says safety is a major part of training at Bechtel Corporation, which boasts a record of zero lost-time accidents at ninety percent of their job sites. "We like to find people who have construction experience so they know how to work safely in that arena. We do a lot of safety training. If you want to lose your job here, the way to lose it is to do anything unsafe. We don't ever want to send anyone home in worse shape than they came to work."

The very nature of welding inherently carries the possibility for injury. According to the Occupational Health and Safety Administration (OSHA): "Health hazards from welding, cutting, and brazing operations include exposures to metal fumes and to ultraviolet (UV) radiation. Safety hazards from these operations include burns, eye damage, electrical shock, cuts, and crushed toes and fingers."

Safety standards were regulated in mid-1943 when wartime industrial demands created a burst of welding activity. Under the auspices of the American Standards Association (ASA), now the American National Standards Institute (ANSI), the first welding standards were officially drafted in 1944. It was titled the *American War Standard Z49.1, Safety in Electric and Gas Welding, and Cutting Operations*. The *Standard* underwent its first revision in 1950, and subsequent revisions occurred in 1958, 1967, 1973, 1983, 1988, 1994, 1999, and 2005. The 2005 revision is a free download at the American Welding Society (AWS), website at www.aws.org. The AWS site also includes fact sheets detailing safety precautions and procedures for a variety of situations from contact lens wear on the job to lockout/tagout procedures to dealing with fumes from toxic metals.

OSHA, the Centers for Disease Control, and Prevention's (CDC) National Institute for Occupational Safety and Health (NIOSH) provide detailed information on welding safety procedures. The illustration on the next page presents information on personal protective equipment for arc welders. It offers a typical overview of safety equipment and concerns – this necessary apparel is all that stands between a welder and the sparks.

Personal Protection

Because of the heat, UV rays, and sparks produced by the arc welder, the operator and his helpers must be attired appropriately. For body protection, wear a pair of fire retardant long-sleeved coveralls without cuffs. Clothing should never have loose pieces, snags, or tears because scattering sparks can easily ignite clothing. Even thin or worn spots in clothes create a hazard. Always keep sleeves and collars buttoned to protect your skin, and hands should be protected with leather gauntlet gloves. High-top leather shoes, preferably safety-rated, should be worn on the feet. If low shoes are worn, fire resistant leggings should protect the ankles. Even though the welder wears a helmet, he should still protect his eyes by wearing transparent goggles if he wears prescription glasses or safety glasses if not. The welding helmet or hand shield with filter plate and cover plate is *mandatory* for eye protection from the harmful rays of the arc. The filter plate should be at least shade No. 10 for general welding up to 200 amps; however, certain operations such as carbon-arc welding and higher current welding operations require darker shades. Never use a helmet if the filter plate or cover lens is cracked or broken. Additionally, many welders wisely use a flameproof skullcap to protect their hair and head. Also, when working in a noisy environment, hearing protection should be worn.

Welders should never carry any ignitable materials in their pockets, especially plastic disposable cigarette lighters. To provide protection to bystanders or other workers, welding should be done inside a properly screened area when possible. When conditions make this impossible, a portable screen can be set up, or bystanders can wear anti-flash goggles.

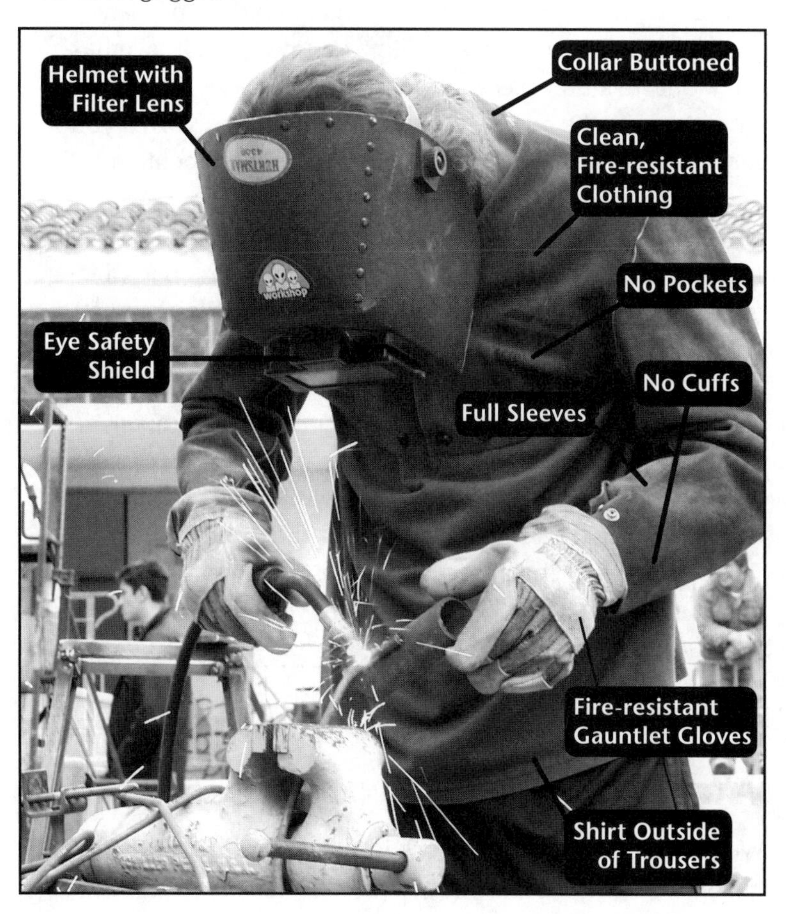

Profile: *Charles Holley, Fluor Corporation*

Like many professional welders, Charles Holley was first introduced to the skill through high school agriculture classes. After high school, he took a construction job with an air conditioning sheet metal company where he performed light welding duties. From there he moved on to Brown and Root, where he began working as a welder's assistant in an Arkansas paper mill. During the three-and-a-half years he worked there, Holley attended the corporation's welding school, where he discovered his life-long career. Holley goes on to explain, "I was welding on the job during the day, and at night, I started attending a local skills school. I learned TIG, MIG, and other skills to help me on my way. In 1985, I went to Fluor and I've been here ever since."

At Fluor Corporation, Holley works as a maintenance mechanic/ welder. "If something on a job site breaks," he says, "I fix it, overhaul it, repair it, fabricate a new piece, or do whatever it takes. I like this job because it's never the same old thing. You never know what you may have to repair or be called on to build. There's no repetition here."

Because of Holley's seniority, he often teaches young welders, a part of his job he finds especially satisfying. "My favorite part of the job is when somebody walks by and says, 'Hey, that looks great! Good job!' I want to instill that kind of pride in the young guys I teach," he says.

He also supervises welding crews at times when the company is taking on certain large-scale projects, but currently has no plans to move into management. At age fifty, Holley is looking toward a future that would include not only official retirement, but also the opening of his own weld shop as a new career. "I may retire from the company someday, but I will never retire from welding as long as my health holds up," he says.

Work Schedules

Welders are likely to find work in whatever hours suit them, especially if they are willing to relocate. There are welding jobs with regular hours and weekends off, with little or no overtime expected. Factory work is likeliest to hold to such hours.

Construction work is another matter. Overtime is common; Dan Barrow says most of Zachry's welders work fifty-hour weeks. Russell Battles says at the Fluor Corporation, four- to eight-week busy seasons can see six- or seven-day work weeks with up to twelve hours of work a day as, for example, Fluor's schedules are dictated per their contracts with the power plants.

Battles explains his division's busy season coincides with the lower-demand season – when milder weather means less need for electricity – for Luminant, Fluor's client. The schedule often entails using temporary welders. "We've got about 500 people regularly," Battles says. "If we go into a scheduled outage, we may ramp up to 1500. A lot of temp welders work outages only. They may work six months a year and work as much as people who work forty hours a week all year along."

Unfortunately, not all Luminant outages are planned. When a unit goes down in the middle of a blazing summer, or a freezing winter for that matter, crews feel extra pressure to get the unit back on line. Battles remembers one such time as being particularly rewarding. "It was back in '96 or '97, a really hot summer, and anytime a boiler would go down for a tube leak, we would go in, make the repairs, and get it back online as fast as possible so they could produce electricity."

For Bechtel welding engineer Jason Praster, the workday begins early, about 6:30 a.m., and ends about ten hours later. Praster explains, "We start with the 'Plan of the Day,' a POD, and everybody goes through what crews are where. We go through start cards and Bechtel protocol, then foremen and crews get together to discuss any special needs of the day and our safety topic for the day. We non-manuals attend to make sure no one is taking shortcuts, and craft [workers] are updated with the proper info for the day." Once the meetings are finished, Praster and the other welding engineers begin their inspection and testing rounds. "There are two penetrant tests," Praster says. "There are two different types – a solvent-removable and a water-washable test. We use an oil-based solvent that penetrates into cracks or crevices within a weld. Then we clean off the oil, and spray developer on it to see if there are any leaks."

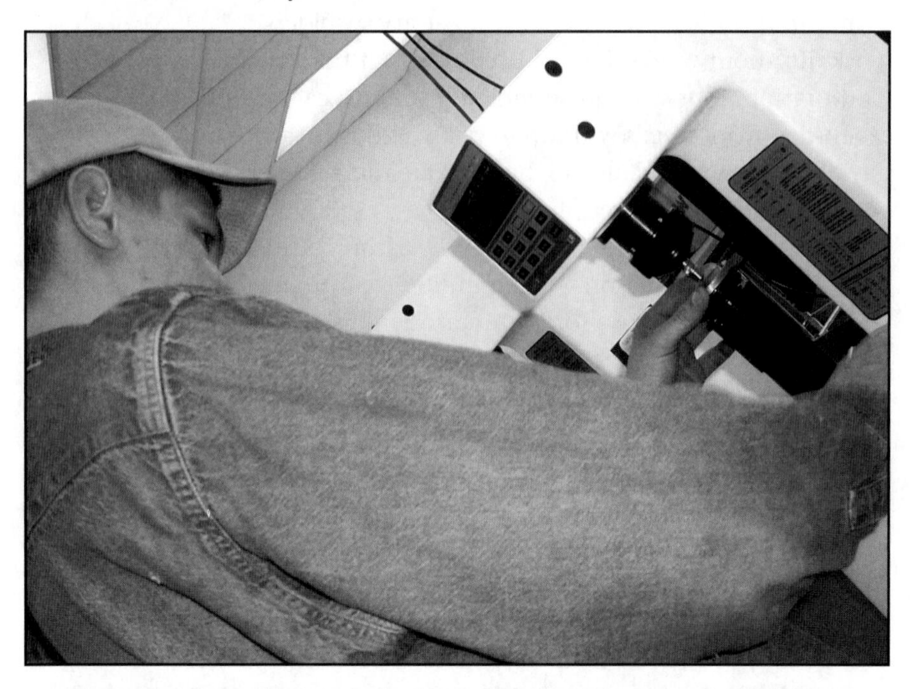

Profile: *Dave Cotner, Pennsylvania College of Technology*

Welding department head Dave Cotner grew up on a farm, where heavily used equipment often broke. He and his father took the damaged implements to a local welder for repair when needed, and even as young as age five, Cotner found himself mesmerized by welding. Everything about it intrigued him. Cotner says, "This guy [the repairman] was a tough little guy, who looked like he could take on the whole world. It made me think this job is tough and cool – it's what I want to do."

As a teenager, Cotner began welding around the house, tinkering with projects. He took the traditional college preparatory route in high school, but also enrolled in a couple of shop classes, where he could hone his manual welding skills.

"The guidance counselors did not pick up on the fact that my interest and skills lay in welding," says Cotner. "Instead, I went into a psychology and counseling major for a while. A few weeks into it, I knew I had made a mistake, and I chose welding." At that point, Cotner dropped out of college and entered the construction field, where he did some welding. Moving to a position at a factory, where he worked as a machinist and welder, he realized he had found his niche. He enrolled in a small welding training school and completed an 840-hour combination course, in which he attended school for four hours a day and worked second shift at a machine shop. Even later, when he worked as a sales service representative for a welding consumables manufacture, he continued to weld as a side job.

At age twenty-four, the newly married Cotner made a deal with his bride. "I told her, the welding thing is where I want to be, but I need a piece of paper behind it," says Cotner. "I had not done well in high school, so I promised her that I would try just one class. We agreed that if I did well, I'd go back to school. I ended up with a Bachelor of Science in welding fabrication and engineering technology at Penn, while working full-time for a bridge girder manufacturer." In 2004, Cotner began teaching full-time at Penn College, and in 2007, he completed his Masters of Education from Wilkes University in Wilkes-Barre, Pennsylvania.

Cotner tells of a student who graduated from Penn College in May 2007. The young man had been raised on a farm in New Jersey and was very nervous about being so far from home. In fact, he made trips home every weekend. During his junior year he began a requirement all welding students in the Bachelor of Science degree program have to meet – senior seminar, which includes a summer internship. Try as he might to find something close to home, this young man ended up going to a Pennsylvania utility company for the summer. "He was nervous about staying here while most the other students were going home for the summer," says Cotner, "so I gave him my cellphone number and told him to call me if he needed me, since I only live a couple of miles from the school. I told him he could hang out with me and my family if he wanted. During the position, he blossomed and grew, and he was eventually offered a full-time position with the company, which he accepted. He did what he thought he could never do. He came here with plans just to get a degree and go home to use welding on the farm. Instead, he got a career."

Prospective students who hope to benefit from the experience and reputation of instructors like Cotner should have drive, desire, a willingness to learn, and some curiosity about the processes involved in welding. "The kids who are excited to be here and want to see something new are the kids we are looking for," says Cotner. The majority of students who go to Penn College have some prior experience in welding, whether on the job or at the high school vocational class level, but Cotner says that in his current class of sixteen students, five had no welding experience at all. He offers this advice to high school students: "Look for opportunities to experience welding first hand. If there is no availability for that, look for a place in your community, someone who is a welder, who will spend time with you. Welders typically like to share their craft. We like to talk about what makes us happy, and welding makes us happy."

He also advises students to focus on math and sciences at the high-school level, as well as blueprint reading, drafting, and Computer-Aided Design (CAD) programs. "Take chemistry and physics. This is a science, and depending on the level you decide to take it to, you may need a lot of science. If you are just arcing and sparking, no, but to be able to perform and calculate some of the functions that are going to show up in physics, strength and materials, statics, calculus, and metallurgy, yes. The foundation you have for something as simple as the construct of an atom will come into play in metallurgy class."

Conclusion

If you have skill and drive as a welder, the jobs are there, the pay is superb, and the challenges are rewarding. Now, more than ever, the future for welders remains as bright as an arc.

Charles Holley, a twenty-three-year veteran at Fluor Corporation, sums it up: "I've seen economic hard times and economic good times, but I have never been without a job when I wanted or needed one. Welding has been a good living for me. My kids have never gone without, and I will weld for as long as I can."

CHAPTER TWO

Welding Education and Certification

Becoming a successful welder starts with a solid education in the skills and knowledge necessary to grow in your career. Some high school students have access to excellent teaching at the high school level through agriculture or shop classes, and some high schools still offer a full vocational technical education with one or more years of welding education.

For the most part, however, welders receive their training in college programs. Potential welders have a choice of degree paths. Technical and community colleges typically offer a choice between a welding diploma – a one-year program – and an associate degree, which is a two-year program. Some universities offer a bachelor's degree in welding engineering, which often combines hands-on welding skills with engineering design and problem-solving skills. Additionally, some universities offer advanced degrees, including a Ph.D. in welding engineering.

In order to know which degree plan would be best for you, consider your personality and academic strengths. For students who are primarily hands-on people without interest in or aptitude for advanced math and sciences, a welding diploma is the fastest way to get from school to the workforce. The associate degree requires some general education classes like English, college algebra, psychology, and/or history. If you are a strong math and science person and as interested in design and problem solving as creating, you might consider the welding engineering program, which will include calculus, physics, chemistry, and related areas.

The National Center for Welding and Training (Weld-Ed), works in collaboration with college and corporations in the advancement of information and understanding about welding careers. Its website, www.weld-ed.org, holds valuable information about the organization's member colleges which offer an education in welding and welding engineering.

Profile: *Frank Wilkins, Texas State Technical College Waco*

Frank Wilkins retired in 2010 as longtime director of the TSTC Waco welding program. He began his welding career in 1966 after he had just come off a tour of duty as a Marine in Vietnam, where he was injured. "I came home to work and my good friend's (Larry Battles) dad hired me with Brown and Root – now Kellogg, Brown and Root," Wilkins remembers. "I practiced my welds in the early morning and at lunch. Pretty soon, Mr. Battles gave me a welding test, and I passed."

Wilkins continued working in the field until 1982 when he accepted a teaching position at TSTC. At the time, he was working on the Comanche Peak nuclear plant in Glen Rose, Texas, commuting ninety-two miles one way to work. "TSTC had an opening, so I went out and applied. I was not even sure I wanted to do it, but I thought I would give it a try," Wilkins says.

As a teacher, Wilkins looked for the same qualities in all students who wanted to be successful in their education and career, regardless of the program a student chose at TSTC. "I want someone who listens, who pays attention, and follows directions; someone who is willing to make new steps. They must be teachable," says Wilkins on teaching. As an educator, Wilkins was passionate about helping students find the right path in life, and his list of high standards didn't not stop there. "Treat the classroom as if it were a job," the former educator says. "Students who do not succeed here are the ones who do not come to class."

The students who followed Wilkins' advice and attended class regularly found an experienced staff well equipped to help students reach their career goals. TSTC has "some of the best faculty members in the world," Wilkins says. "They have been out there; they've done it. They're excellent teachers, and they know what's going on."

To further students' chances of success and give them better access to faculty, TSTC keeps their classes small, below the state minimum when possible. Wilkins says, "The state of Texas allows twenty students in a welding class, but we did about fifteen. They get that one-on-one that is so important that way." Of the approximately fifteen students in a class, five students were women. Statistically nationwide, only about five percent of professional welders are female. "We really worked on eliminating gender bias in the class," Wilkins says, "but in the work world, getting past the male chauvinist is the biggest hurdle. Most female welders I have seen are very good, but girls need to know there is a bias against females in the field."

Like most welding instructors and employers, Wilkins is very concerned with the welder shortage rates. TSTC works closely with Weld-Ed to get the word out to high school teachers in Texas about the critical need for skilled welders. "Weld-Ed is designed to get counselors and high school teachers interested in welding and make them understand the shortages. We are doing a pretty good job here in Texas," Wilkins says. "More high school teachers are realizing this and relaying the message, but plenty of schools still don't teach it. The older welders are retiring. Who is coming up behind them? Thirty-four new nuclear plants are going to be built in the next few years. We will have a tremendous shortage and have to import welders from overseas. Many big companies already are bringing in welders from China, Puerto Rico, Mexico, and India."

In his former position as department chairperson at TSTC, Wilkins took his role in recruiting new students seriously. He would spend time in the high schools, discussing educational, and job opportunities with young prospective welders. He felt it was important for students to understand that welding requires more skills than just good hand-eye coordination. "I actually had a counselor tell me welding students do not need reading, writing, or math skills," he says, incredulously. "Typically, welders *are* outdoors types; people who like to design then build things with their hands; people who don't mind heat. But to deal with fabrication, you need to know math," Wilkins insists. "Pipe fabricators need to know welding, plus algebra and trigonometry. Welding engineers need all that and calculus and physics. They are studying types of materials to be used in different service conditions and learning thermodynamics, chemical reactions, and metallurgy."

Despite retirement, Wilkins is still a man with a passion for a career that can change the lives of students. "I am helping to get the word out to the high schools and counselors about this career – about its opportunities and what is expected of a welding technician. It's hot and dirty sometimes, but not always. Welding does require reading, writing, and math. Jobs are there, the needs are there, and they are going to be there for a long time. These are very well-paid careers that can take you as far as you want to go in life. TSTC is a great place to get started."

Educational Requirements

Bob Alexander says some have never seen a welding machine before coming for training. Holding a degree prior to being hired, however, can mean starting in positions of higher pay and more responsibility. "The Associate of Science degree can open doors for welding Quality Control (QC), and welding engineering opens even more opportunity because you get more training than just the act of welding. We hire based on resume and specific area of training for specific job sites' needs," Alexander says.

College plans run from a one-year certificate program to the two-year associate degree (usually Associate of Applied Science) to a four-year bachelor's degree in welding engineering. The certificate is sufficient for an entry-level job, but better-paying welding technician jobs require at least an associate degree (or substantial welding experience), and welding engineering jobs absolutely require at least a bachelor's degree (at present only The Ohio State University offers master's and Ph.D. programs).

The Certificate and Two-Year Degree

Certificate programs typically offer a bare-bones introduction to welding techniques, just enough to get you in the door with an employer. The two-year associate degree – generally around seventy hours of coursework – provides a more well-rounded education that includes deeper study of metallurgy and computer-based welding equipment, as well as courses in math, communication, psychology, and business.

Frank Wilkins estimates up to eighty percent of his students came to TSTC with some basic knowledge of welding skills, but this is not always the case. For this reason, whether a student picks the certificate program or the degree program, the first year of school looks the same for all welding students. That first year focuses completely on skill development, learning how to handle the equipment and how to make a solid, clean weld. The students take diverging paths the second year. "The two-year students go on to find out how things work, and the certificate students go out to work in the field," said Wilkins. "Once the welding technicians graduate with their Associate of Science degree, they have options that certificate students would not have. They may work as inspectors or as junior engineers. They could go on to a four-year university for a bachelor's degree in welding engineering. They can't do specialties like automation and metallurgy."

Profile: *Nicole Lankford, TSTC Waco*

Nicole Lankford was one of the few female graduates of TSTC in the spring of 2008. At age twenty-eight, she was a career-changer. Lankford had decided that with two small children to feed, a secretary's salary was not going to get her where she wanted to be in life. Lankford says, "I was looking for a change. I was working in an office for $12 an hour with two kids, and there just wasn't enough [money]. So, I started looking around for a school I could finish quickly and a career that made more money. As I checked things out, welding sounded interesting, and the money was great, so that caught my attention."

Knowing that TSTC offered two different welding programs, Lankford chose the associate degree program because of the opportunities that it would offer later in life. "If all you want to do is weld, then the certificate program is for you. If you want to be in charge some day, then you need to go for the degree program."

Lankford found TSTC's instructors to be knowledgeable and easy to learn from. Although she found tungsten inert gas (TIG) welding to be challenging at first, with persistence and help from her teachers, she quickly caught on. "If you have a problem with something, talk to your instructors about it," she says. "If you cannot do something, they will throw on a hood and show you how it's done." Lankford also appreciated the help she received when it came time for bookwork. Because the degree program offers classes like metallurgy, many students find they need more help in the advanced classes. "Welding has so many different aspects; it was nice to be able to dabble in all of them and get help when I needed it. The instructors always were willing and ready to explain new things."

Lankford describes herself as a hands-on person, one who chafes if stagnant for too long. When asked to describe her favorite part of her education at TSTC, her answer is simple: "Getting in there and welding. I like to watch hot metal. I like the way it smells and the way it looks. I like playing with fire and metal and seeing what I can make out of them," says Lankford, who plans to weld artistically in the future.

Lankford's passion for hot metal already has landed her a great job with Kiewit Corporation as a welding engineer. Prior to moving her family to the Gulf Coast of Texas to begin her new career, she finished a pipe-welding certificate on top of her associate degree. Once on the job, Lankford will be trained to perform procedures and write them up for the welders. Although she will not be welding on a daily basis, she is glad she will have regular office hours, which will allow more time for her to be with her children. As one who likes to be on the move, Lankford is happy because Kiewit works on topsides for oil rigs, even though she will spend plenty of time in the shop.

Does she expect to face discrimination in the shop? "I definitely expect to face discrimination," she says. "It's a man's world. Most of the women there are secretaries. Some things are difficult, like fitting in with everyday conversation. When it comes to heavy lifting, men have the advantage, but when it comes to welding, it's not more difficult for a woman than a man."

Despite the gender obstacles and biases Lankford expects to face, she is confident in her abilities. "I know it will be a little difficult to adjust, but I can handle it. I already have been doing it for two years – nothing but me and the hood. TSTC has done a good job to prepare me."

Welder salaries start in the ballpark of $45,000 per year. Lankford happily recommends the degree program at TSTC to anyone interested in becoming a welder, male or female. "There is no reason a woman should not get in there and earn a man's salary. If this is the career you want, TSTC is the place to go. They will help you get started, learn your skills, and get a job."

The Four-Year Bachelor's Degree

The bachelor's degree takes welding students into the realm of welding engineering, teaching the design and engineering principles behind the welded structures. The degree program at The Ohio State University (OSU), explains associate professor Dr. Charles Albright, looks very similar in curriculum to material sciences or mechanical engineering. "There are four areas associated with this degree," says Albright. "There's the design part, which employs mechanical engineering. The materials part is comprised mostly of practical metallurgy. Then there's the electrical control that is comprised of electrical engineering principles. There are also a number of physics processes, especially in non-destructive testing, including X-ray, ultrasonic inspection, etc., so the student must be familiar with physics engineering, too."

To help students gain an understanding of manual skills integrated with engineering, all incoming students take two welding labs to learn or hone manual skill. However, this program is mainly engineering oriented, according to Albright. "Here we are most concerned about process control, metallurgy, and welding design. We are very academic. Most welding done in other parts of the program are robotic or machine welding. If someone comes in with really good manual skills, though, it sets them apart. There are some engineers who work closely with highly trained manual-skills welders. You have to know what the machine should be doing as it simulates a human welder."

Profile: *Dr. Charles Albright, OSU*

Dr. Charles Albright came to OSU in 1979. His educational resume is impressive, with a bachelor's degree in metallurgic engineering from Michigan Technological University, work in nuclear engineering from Northwestern University, and a Ph.D. in material sciences.

"I'm a metallurgist who became involved in welding and joining," says Albright. Sometimes, OSU gets welders who desire to become engineers. "If we get a welder from a two-year technical school, and he wants to be an engineer, he usually has to take classes in chemistry, physics, and calculus. It's not a continuous career path," he says, "but we get a number of people each year who decide to bite the bullet and do it. Most two-year colleges are also teaching chemistry, and some teach physics and calculus. If it is equivalent to our programs, we may be able to transfer the credit."

An interesting development in welding is the use of friction stir, which students at OSU's program learn. "Friction stir is being used in many applications, including rocket engines, fuel tanks, and high performance auto rims made out of aluminum," says Albright. "With friction stir, a tool resembling a coarse screw is used to deform the surfaces being joined, essentially stirring the metals together in the solid state; the interfaces [joints] are completely eliminated. It is a remarkable process."

High school students interested in welding engineering would do well to prepare for the rigorous engineering standards by taking classes such as advanced placement (AP) physics, chemistry, calculus, and any pre-engineering courses available. The program at OSU definitely has a selective admissions process, and students who have prepared at the high school level will be more likely to gain admission.

If, however, a student does not come from a solid math/science background at the high school level, Albright offers a final piece of advice to anyone interested in this interesting and lucrative career: "In terms of academic abilities, most people are smarter at twenty-one or twenty-two than they were at seventeen or eighteen. Do not give up on your academic learning potential. It gets easier as you get older, even if someone was not such a great student in high school because they were distracted or unmotivated. Try some classes you think you may not like; give yourself a chance. This is a great career."

Graduate Degrees

The pioneering welding engineering program at OSU is currently the only one in the nation to offer advanced degrees in welding, with both a master's and a Ph.D. available. OSU also offers an accelerated bachelor's/master's program that allows undergraduates to complete some of the requirements for a future master's degree while pursuing their bachelor's degree.

AWS Certifications

Recognized as the leading authority for certification and continuing education in welding careers, the AWS was founded in 1919. This multifaceted, nonprofit organization has published the following mission statement:

The mission of the American Welding Society is to advance the science, technology, and application of welding and allied joining and cutting processes, including brazing, soldering, and thermal spraying.

In keeping with these goals, the AWS offers certification and continuing education to all qualified welders. You can find the AWS on the Web at www.aws.org.

The following are links and information regarding the various welding certifications available through the AWS:

Certified Welder (www.aws.org/w/a/certification/CW/): The Certified Welder (CW) program is a performance-based program with no prerequisite courses or certifications required. Final certification will provide "transferable" credentials that you may take with you wherever you go. The CW program tests procedures used in the structural steel, petroleum pipelines, sheet metal, and chemical refinery welding industries.

Certified Welding Inspector (www.aws.org/w/a/certification/CWI/): The Certified Welding Inspector (CWI) certification is widely recognized, both nationally and internationally, in the welding industry. Successful companies have come to rely on this AWS certification when ensuring the highest level of quality workmanship.

Certified Associate Welding Inspector (www.aws.org/w/a/certification/CAWI/): As a Certified Associate Welding Inspector (CAWI), a welder gains hands-on experience working alongside a Senior Certified Welding Inspector

(SCWI) or Certified Welding Inspector (CWI). The decision to become a CAWI signals employers that you're serious about your career.

Senior Certified Welding Inspector (www.aws.org/w/a/certification/SCWI/): A welder who has been a CWI for a minimum of six years and has evolved his or her career into supervisory or managerial responsibilities in the field of quality control/quality assurance may apply to become a SCWI, the highest level of certification. The SCWI is an individual who is in a supervisory position and has attained a higher level of experience than a CAWI. The successful applicant also has professional education, which has provided the opportunity to solve problems with a scope and level of difficulty beyond those experienced as a CAWI.

Certified Welding Educator (www.aws.org/w/a/certification/CWE/): The Certified Welding Educator Program (CWE) is geared for the welding professional specifically in the welding education field. This AWS certification confirms ability, talent, and knowledge to specifically direct and perform operations associated with welder training and classroom instruction.

Certified Radiographic Interpreter (www.aws.org/w/a/certification/CRI/): The holders of this certification will have a valuable tool to demonstrate their qualifications to interpret radiographs of weldments. The new AWS program for Certified Radiographic Interpreters (CRI) will be available only to those individuals who successfully pass the required examinations.

Certified Welding Supervisor (www.aws.org/w/a/certification/CWS/): The Certified Welding Supervisor (CWS) program offers welding supervisors and their companies the opportunity to put the welding supervisor in a support position for the welders, making them the most productive and best they can be. This innovative program identifies a body of knowledge all welding supervisors should know and understand in order to increase productivity and improve weld quality.

Certified Welding Engineer (www.aws.org/w/a/certification/CWEng/): The Certified Welding Engineer (CWEng) is capable of directing those operations associated with weldments and other types of joints that are completed in accordance with the appropriate contract documents, codes, and other standards to produce a satisfactory product. The welding engineer's activities begin before production or construction welding and continue through the production process, ending when the production process is complete.

Certified Robotic Arc Welding (www.aws.org/w/a/certification/CRAW/): The Certified Robotic Arc Welding (CRAW) certification allows many welding personnel employed in various welding sectors to measure themselves against standards for their occupation. It also signifies that the CRAW Operator or the technician has demonstrated the capability of working with various codes, standards, and specifications. Because proof of active practice or re-examination is required every three years, certification also signifies that the CRAW Operator or the technician is current with the welding industry.

Certified Welding Fabricator (www.aws.org/w/a/certification/FAB/): The Certified Welding Fabricator (CWF) is responsible for control of contract documents and procedures, material control, welding, inspection, and shipment. The welding fabricator is required to operate an internal quality control program. An appropriate welding quality system is the foundation of delivering a quality welded product or service. When designed for the welding fabricator's unique products and suitably committed to paper and practice, the daily manufacturing operations of the welding fabricator are more consistent and traceable when problems arise.

Conclusion

While welders can still enter the field with only a high school education, the days of welding as an entirely low-tech job are long gone – making college education a key to advancing in a welding career.

"There's a misconception about the welding industry that we are in the lower end of the educational system," says Joel Johnson, coordinator of North Dakota State College of Science's manufacturing and welding program. "There is so much applied physics and chemistry that go into welding. If you are going to be a good welder or welding tech, you have to be more educated to succeed in the industry today. We use fifty-five elements in the periodic table just in daily welding. Students who are good at applied sciences are usually excellent welding students."

Notes

CHAPTER THREE

Additional Welding Information and Resources

Higher Education Welding Programs in the United States

ARKANSAS

Pulaski Technical College
3000 W. Scenic Dr.
North Little Rock, AR 72118
501.812.2200
www.pulaskitech.edu/programs_of_study/welding/default.asp

CALIFORNIA

Fullerton College
321 E. Chapman Ave.
Fullerton, CA 92832
714.992.7484
www.fullcoll.edu

Mt. San Antonio College
1100 N. Grand Ave.
Walnut, CA 91789
909.274.7500
www.mtsac.edu

Pasadena City College
1570 E. Colorado Blvd.
Pasadena, CA 91106
626.585.7123
www.pasadena.edu

Shasta College
11555 Old Oregon Trail
Redding, CA 96049
530.242.7500
www.shastacollege.edu/welding

Solano Community College
4000 Suisun Valley Rd.
Fairfield, CA 94534
707.864.7000
www.solano.edu

GEORGIA

Savannah Technical College
5717 White Bluff Rd.
Savannah, GA 31405
912.443.5700
www.savannahtech.edu

IDAHO

Idaho State University
921 S. 8th Ave.
Pocatello, ID 83209
208.282.0211
www.isu.edu/ctech/welding

ILLINOIS

Illinois Central College
1 College Dr.
East Peoria, IL 61635
309.694.5422
www.icc.edu/ait

Kankakee Community College

100 College Dr.
Kankakee, IL 60901
815.802.8100
www.kcc.edu

Kaskaskia College

27210 College Rd.
Centralia, IL 62801
618.545.3406
www.kaskaskia.edu/Welding

KANSAS

Cowley College

125 S. 2nd St.
Arkansas City, KS 67005
800.593.2222
620.442.0430
www.cowley.edu/departments/it/welding/

Johnson County Community College

12345 College Blvd.
Overland Park, KS 66210
913.469.8500
www.jccc.edu

Kansas City Kansas Community College

Technical Education Center
2220 N. 59th St.
Kansas City, KS 66104
913.627.4100
www.kckcc.edu/academics/programsDegrees/
weldingTechnologyTecCert.aspx

KENTUCKY

Southeast Kentucky Community and Technical College

700 College Rd.
Cumberland, KY 40823
606.589.2145
southeast.kctcs.edu

Owensboro Community and Technical College

4800 New Hartford Rd.
Owensboro, KY 42303
270.686.4400
866.755.6282
www.owensboro.kctcs.edu

MICHIGAN

Monroe County Community College

1555 S. Raisinville Rd.
Monroe, MI 48161
734.242.7300
www.monroeccc.edu/industrial/welding_tech.htm

Muskegon Community College

221 S. Quarterline Rd.
Muskegon, MI 49442
231.773.9131
866.711.4622
www.muskegoncc.edu

West Shore Community College

3000 N. Stiles Rd.
Scottville, MI 49454
231.845.6211
www.westshore.edu

MINNESOTA

Saint Paul College

235 Marshall Ave.
St. Paul, MN 55102
800.227.6029
651.846.1600
www.saintpaul.edu/programs/Pages/Welding.aspx

MISSOURI

Hillyard Technical Center

3434 Faraon St.
St. Joseph, MO 64506
816.671.4170
www.hillyardtech.com/departments.
php?area=weldingtechnology

Metropolitan Community College

3200 Broadway
Kansas City, MO 64111
816.604.1000
mcckc.edu/mccbtc/TrainingWelding.asp

MONTANA

Montana State University - Great Falls

2100 16th Ave. S.
Great Falls, MT 59405
800.446.2698
406.771.4300
www.msugf.edu

Montana Tech of the University of Montana

1300 W. Park St.
Butte, MT 59701
406.496.4101
www.mtech.edu

NEBRASKA

Southeast Community College
8800 "O" St.
Lincoln, NE 68520
800.642.4075
402.471.3333
www.southeast.edu

NEW MEXICO

Doña Ana Community College of New Mexico State University
3400 S. Espina St.
Las Cruces, NM 88003
800.903.7503
505.527.7500
dacc.nmsu.edu/tsd/welding

New Mexico Junior College
1 Thunderbird Cir.
Hobbs, NM 88240
505.492.2859
www.nmjc.edu/academics/programs/welding.asp

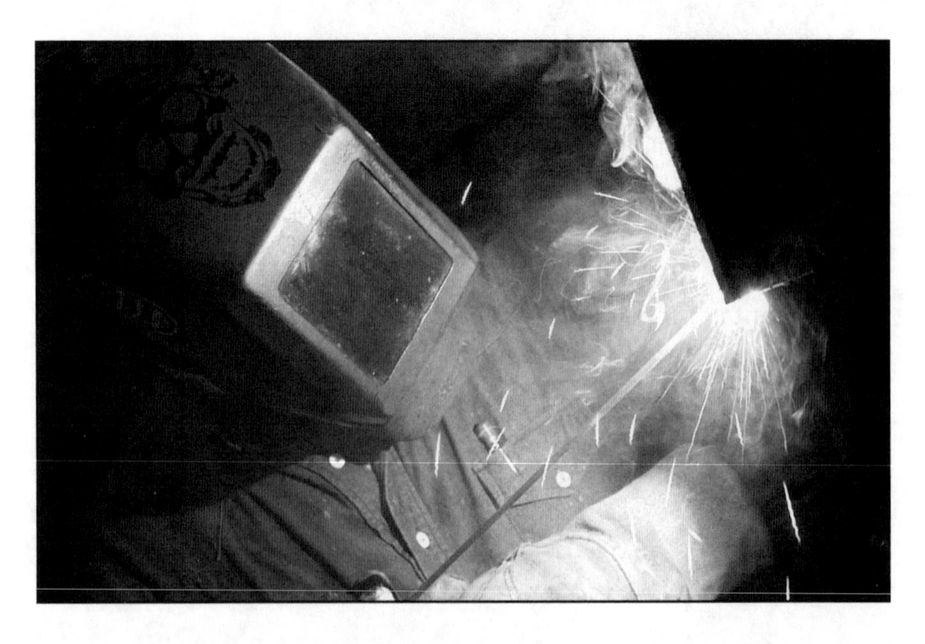

NORTH CAROLINA

Isothermal Community College

286 ICC Loop Rd.
Spindale, NC 28160
828.286.3636
www.isothermal.edu/appsci/welding/welding%20
home.htm

Stanly Community College

141 College Dr.
Albemarle, NC 28001
704.982.0121
www.stanly.edu

NORTH DAKOTA

Bismarck State College

1500 Edwards Ave.
Bismarck, ND 58506
800.445.5073
701.224.5400
www.bismarckstate.edu

North Dakota State College of Science

800 Sixth St. N.
Wahpeton, ND 58076
800.342.4325
ndscs.admissions@ndscs.edu
www.ndscs.edu

OHIO

Lorain County Community College

1005 N. Abbe Rd.
Elyria, OH 44035
800.995.5222
www.lorainccc.edu

The Ohio State University – College of Engineering
477 Watts Hall
2041 College Rd.
Columbus, OH 43210
614.292.2553
www.matsceng.ohio-state.edu/weldingengineering

OREGON

Lewis & Clark College
0615 S.W. Palatine Hill Rd.
Portland, OR 97219
503.768.7000
www.lclark.edu

Southwestern Oregon Community College
1988 Newmark Ave.
Coos Bay, OR 97420
800.962.2838
541.888.2525
www.socc.edu/academics/pgs/academic-dept/fabrication-welding/index.shtml

Umpqua Community College
1140 Umpqua College Rd.
Roseburg, OR 97470
541.440.4600
www.umpqua.edu/Welding

PENNSYLVANIA

Northampton Community College
3835 Green Pond Rd.
Bethlehem, PA 18020
610.861.5300
www.northampton.edu

Pennsylvania College of Technology

1 College Ave.
Williamsport, PA 17701
800.367.9222
570.326.3761
www.pct.edu/schools/IET/weld

SOUTH CAROLINA

Greenville Technical College

620 S. Pleasantburg Dr.
Greenville, SC 29607
864.250.8111
www.gvltec.edu/technology.aspx?id=926

Northeastern Technical College

1201 Chesterfield Hwy.
Cheraw, SC 29520
800.921.7399
843.921.6900
www.netc.edu

Piedmont Technical College

620 N. Emerald Rd.
Greenwood, SC 29648
800.868.5528
www.ptc.edu/academics/areas-of-studying/welding

TENNESSEE

Tennessee Technical Center Crossville

910 Miller Ave.
Crossville, TN 38555
877.811.7502
931.484.7502
www.ttcc.edu

Tennessee Technical Center
1100 Liberty St.
Knoxville, TN 37919
865.546.5567
www.ttcknoxville.edu

TEXAS

Central Texas College
6200 W. Central Texas Expwy.
Killeen, TX 76549
800.223.4760 (in TX)
800.792.3348 (out of TX)
254.526.1249
www.ctcd.edu/industech/ind_wldg.htm

Hill College
112 Lamar
Hillsboro, TX 76645
254.659.7500
www.hillcollege.edu

Houston Community College
3100 Main St.
Houston, TX 77002
713.718.2000
www.hccs.edu

South Plains College
1401 S. College Ave.
Levelland, TX 79336
806.894.9611
www3.southplainscollege.edu

Tarrant County College
1500 Houston St.
Fort Worth, TX 76102
817.515.8223
www.tccd.edu/Courses_and_Programs/Program_
Offerings/Welding.html

Texas State Technical College Harlingen
1902 N. Loop 499
Harlingen, TX 78550
956.364.4814
www.harlingen.tstc.edu/welding/

Texas State Technical College Marshall
2650 East End Blvd. S.
Marshall, TX 75671
903.923.3303
www.marshall.tstc.edu/areas/welding

Texas State Technical College Waco
3801 Campus Dr.
Waco, TX 76705
254.867.4884
www.waco.tstc.edu/wlt

Texas State Technical College West Texas
305 Booker Ave.
Brownwood, TX 76801
325.643.5987
www.westtexas.tstc.edu

Vernon College
4400 College Dr.
Vernon, TX 76384
940.552.6291
www.vernoncollege.edu

The Victoria College
2200 E. Red River St.
Victoria, TX 77901
877.843.4369
361.573.3291
www.victoriacollege.edu/welding

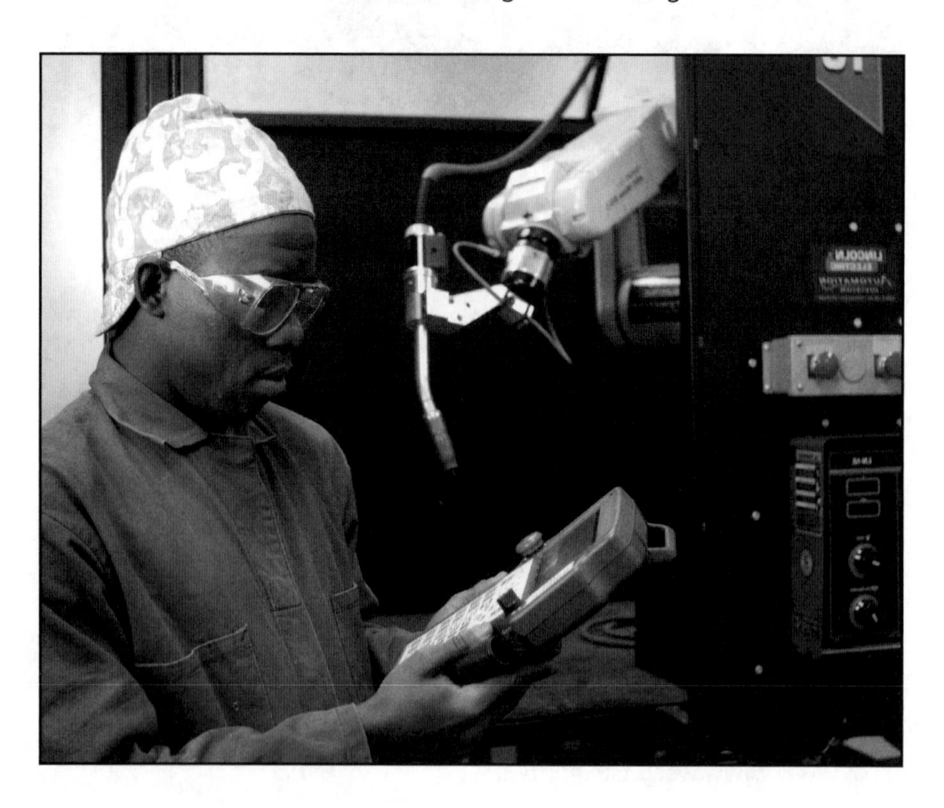

UTAH

Bridgerland Applied Technology College
1301 N. 600 W.
Logan, UT 84312
866.701.1100
www.batc.edu

College of Eastern Utah
451 E. 400 N.
Price, UT 84501
435.613.5000
www.ceu.edu/departments/weld/default.aspx

Weber State University
1802 University Cir.
Ogden, UT 84408
801.626.6305
www.weber.edu/COAST/welding.html

WASHINGTON

Everett Community College
2000 Tower St.
Everett, WA 98201
425.388.9100
www.everettcc.edu/programs/bwe/
advancedmanufacturing/welding_fabrication

South Seattle Community College
6000 16th Ave. SW
Seattle, WA 98106
206.764.2882
www.southseattle.edu/programs/proftech/weldfab.htm

WISCONSIN

Chippewa Valley Technical College
620 W. Clairemont Ave.
Eau Claire, WI 54701
800.547.2882
715.833.6470
www.cvtc.edu

Moraine Park Technical College

700 Gould St.
Beaver Dam, WI 53916
920.887.4490
www.morainepark.edu

Northcentral Technical College

1000 W. Campus Dr.
Wausau, WI 54401
715.675.3331
www.ntc.edu

Waukesha County Technical College

800 Main St.
Pewaukee, WI 53072
262.691.5566
www.wctc.edu/programs_&_courses/skilled_trades/metal_fabrication_welding

WYOMING

Eastern Wyoming College

3200 W. C St.
Torrington, WY 82240
866.327.8996
www.ewc.wy.edu/programs/departments/welding.cfm

Sheridan College

3059 Coffeen Ave.
Sheridan, WY 82801
800.913.9139
307.674.6446
www.sheridan.edu

Welding Two- and Four-Year Degree Plans

Associate of Applied Science Degree in Welding Technology Texas State Technical College Waco

The welding programs at Texas State Technical College (TSTC) emphasize the development of real, hands-on welding, layout and fitting skills. With extensive exposure to welding practices and principles, students can better understand not only how welding processes work, but also why certain welding processes are used. In addition to 180 ventilated arc welding stations and 20 oxy-acetylene stations, Welding Technology offers Combination Welding students instruction on plasma torches for oxy-acetylene and air carbon arc cutting. Students also gain extensive skills and knowledge through simulated industrial welder qualification tests with the following processes: shielded metal arc welding (SMAW), gas metal arc welding (GMAW), gas and self-shielded flux-cored arc welding, gas tungsten arc welding (GTAW), and submerged arc welding (SAW).

	TSTC Waco: Welding Technology	
Students Starting Fall 2010	Associate of Applied Science Degree	Total Credits: 72
Semester One		
WLDG 1313	Introduction to Blueprint Reading for Welders	3
WLDG 1428	Introduction to Shielded Metal Arc Welding	4
WLDG 1430	Introduction to Gas Metal Arc Welding	4
ENG 1301	Composition I	3
	Semester Total:	**14**
Semester Two		
WLDG 1312	Introduction to Flux Core Welding (FCAW)	3
WLDG 1310	Liquid Penetrant/Magnetic Particle Testing	3
WLDG 1417	Introduction to Layout and Fabrication	4
WLDG 1457	Intermediate Shielded Metal Arc Welding (SMAW)	4
ACGM X3XX	Gen Ed Elective	3
	Semester Total:	**17**
Semester Three		
WLDG 1312	Introduction to Gas Tungsten Arc Welding (GTAW)	4
WLDG 1435	Introduction to Pipe Welding	4
WLDG 2443	Advanced Shielded Metal Arc Welding (SMAW)	4
ACGM X3XX	Gen Ed Humanities/Fine Arts Course	3
	Semester Total:	**15**
Semester Four		
ACGM X3XX	Gen Ed Math/Natural Sciences Course	3
WLDG 1337	Introduction to Metallurgy	3
WLDG 2451	Advanced Gas Tungsten Arc (TIG) Welding	4
WLDG 2413	Intermediate Welding Using Multiple Processes	4
	Semester Total:	**14**

Semester Five		
NDTE 1310	Liquid Penetrant/Magnetic Particle Testing	3
WLDG 2350*	Orbital Tube Welding	3
WLDG 2355	Advanced Welding Metallurgy	3
ACGM X3XX	Gen Ed Social Science Course	3
	Semester Total:	**12**

*Capstone course: A required learning experience that results in a consolidation and synthesis of a student's educational experience. The capstone experience certifies mastery of entry-level work place competencies.

Associate in Applied Science in Welding Technology North Dakota State College of Science

The welding curriculum is designed to provide students experience in welding as it pertains to assembly, manufacturing, energy, and construction. This program provides education and training in common welding processes, robotics, Computer Numerical Control (CNC)

cutting, CNC press brake operations, inspection, print reading, fabrication, pipe and plate welding, automated manufacturing, human relations, communications, and other aspects of general education. Career opportunities offer a wide range of employment possibilities in the manufacturing, steel construction, mining, energy, petroleum, and other production areas. North Dakota State College of Science (NDSCS) is an American Welding Society SENSE certified facility. AWS Level I and Level II certification is available.

	North Dakota State College of Science	
	Certificate	
MFGT 101	Robotics I	2
MFGT 123	Fabrication Methods I	2

MFGT 135	Basic Metallurgy	2
MFGT 137	Print Reading I	2
MFGT 151	Welding Theory I	3
MFGT 152	Welding Theory II	3
MFGT 153	Welding Lab I	5
MFGT 154	Welding Lab II	5

	North Dakota State College of Science	
	Applied Associate Degree in Applied Science	
MFGT 101	Robotics I	2
MFGT 123	Fabrication Methods I	2
MFGT 135	Basic Metallurgy	2
MFGT 137	Print Reading I	2
MFGT 140	Fabrication Methods II	3
MFGT 141	Print Reading II	3
WELD 151	Welding Theory I	3
WELD 152	Welding Theory II	3
WELD 153	Welding Lab I	5
WELD 154	Welding Lab II	5
WELD 201	Welding Theory III	4
WELD 202	Welding Theory IV	4
WELD 211	Welding Lab III	7
WELD 212	IV/Pipe/Plate or Welding Lab	7
WELD 213	IV/Fabrication	7

	North Dakota State College of Science Related/General Education Courses	
	Certificate	
CIS 101	Computer Literacy	2
MATH 130	Technical Mathematics	2
PSYC 100	Human Relations In Organizations	2

	Diploma	
CIS 101	Computer Literacy	2
ENGL 105 or ENGL 110	Technical Communications	3
	College Composition I	
MATH 130	Technical Mathematics	2
MATH 132	Technical Algebra I	2
	Social and Behavioral Sciences elective (ECON, HIST, POLS, SOC/SOCI, GEOG)	2
	Wellness elective	1
	Associate	
CIS 101	Computer Literacy	2
ENGL 110	College Composition I	3
English/Communication elective (choose one)		
ENGL 105	Technical Communications	3
ENGL 120	College Composition II	
ENGL 125	Intro. To Professional Writing	
COMM 110	Fundamentals of Public Speaking	
MATH 130	Technical Mathematics	2
MATH 132	Technical Algebra I	2
MATH 136	Technical Trigonometry	2
	Social and Behavioral Sciences elective (ECON, HIST, POLS, SOC/SOCI, GEOG)	2
	Wellness electives	2
	Total Required for Certificate:	**30**
	Total Required for Diploma:	**62**
	Total Required for Associate:	**68**

	North Dakota State College of Science	
	Associate Degree in Applied Science	Total Credits: 69
Semester One		
WELD 153	Welding Lab I	5

WELD 151	Welding Theory I	3
MATH 130	Technical Mathematics	2
MFGT 137	Print Reading I	2
CIS 101	Computer Literacy	2
MFGT 123	Fabrication Methods I	2
FYE 101	Science of Success	1
	Semester Total:	**17**
Semester Two		
WELD 154	Welding Lab II	5
WELD 152	Welding Theory II	3
MFGT 135	Basic Metallurgy	2
ENGL	English elective	3
MFGT 101	Robotics I	2
PSYC 100	Human Relations In Organizations	2
	Semester Total:	**17**
Semester Three		
WELD 211	Welding Lab III	7
WELD 201	Welding Theory III	4
MFGT 140	Fabrication Methods II	2
MATH 132	Technical Algebra I	2
ENGL	English elective	3
	Semester Total:	**18**
Semester Four		
WELD 202	Welding Lab IV	4
MFGT 141	Print Reading II	2
WELD 212 or 213	IVPipe/Plate or Welding Lab IV/Fabrication	7
MATH 136	Technical Trigonometry	2
	Wellness Elective	2
	Semester Total:	**17**

Profile: *Joel Johnson, NDSCS*

As a high school student, Joel Johnson, program coordinator of NDSCS's Welding Program, had the opportunity to learn manual skills welding. After high school, he earned a certificate in welding from the Area Vocational Technical Institute in Thief River Falls, Minnesota. Because his welding classes from high school articulated to the college, he was able to complete his certificate in only one year. The day he graduated, he got a job. "That was harder to do in the early '80s," says Johnson. "There were lots of welders back then because vocational education was huge in the U.S. through the '70s and '80s. That's why we're feeling the repercussions now. Those baby boomers that were graduating from Voc-Ed classes then are retiring now, and we don't have enough skilled people to replace them."

Johnson eventually resumed his education, earning his Certified Welding Inspector (CWI) certification from the American Welding Society (AWS), as well as a bachelor's degree in vocational education from Valley City State University and a master's degree in career and technical education from the University of North Dakota. Since 1993, he has been on staff as a welding instructor at NDSCS, one of the few community colleges in the country with a welding program that is more than seventy years old.

NDSCS's welding students have a choice of whether to earn a certificate, a diploma, or an associate degree. Most of the students who come to the college have just graduated high school, and many, but not all, have previous welding experience. The majority of students choose the degree program, understanding that management and promotions in the career require degrees to back manual skills. In looking for new students, Johnson likes to see people with an inclination toward mechanical skills. "We used to get a lot of kids off the farm, and they were mechanically sound, as far as how to fix things. We just had to teach them to weld. Now, most students do not know how to assemble hands-on projects anymore. We have to start with the mechanics, then move to welding."

Although all students take the same classes during their first year at NDSCS, the second year of the degree program introduces manufacturing processes. Johnson says "We have the power plant construction industries around here, but we revolve mostly around manufacturing, and we physically build projects from scratch." In fact, the welding department has an annual contract with the North Dakota Game and Fish Department to build boat docks for public use. "We originate the plans, students order the materials, and make jigs and fixtures. An inspector makes sure the welding is up to par. Sportspeople throughout the state use those docks. We build ten to twelve of the projects each year. It's a great example of government agencies working together."

This year, the class may be taking on an even more challenging project, as the department has a second-year student who is paraplegic. "We are going to try to design a stand-up wheelchair for him, so he can reach the equipment. If we can do it, the students will design and build it, supervised by the staff and myself," says Johnson.

Johnson sees the future of welding as becoming increasingly automated and more demanding in the way of credentials. He says, "We already have robotic cells here on campus. The other big trend is certification of welders. Even the guys in the construction field or those who are welding for railroad car repair need to be certified. I am seeing more and more companies wanting their welders to have certification. We use the AWS codes. Companies are looking increasingly for someone who meets those standards."

For the welder who meets high standards and has a solid work ethic, Johnson believes the sky is the limit, and he holds himself up to his students as an example. "You can go anywhere you want and do anything you want with this career," he asserts. "I can do things a normal person cannot do – whether that's welding for a hobby or for a job, or learning about it to be a better teacher, or to say I'm a welder with a master's degree. Academically, I was not that good in high school, but I have made it to a higher level. I learned my craft as just a regular shop welder for thirteen years. I bring that to my teaching. You never know where you're going to be when you choose welding for a career. There aren't too many master-degreed welders out there."

Welding Certificate, Associate in Applied Science in Welding Technology, and Bachelor of Science in Welding and Fabrication Engineering Technology Pennsylvania College of Technology - Williamsport, Pennsylvania

Pennsylvania College of Technology's Welding Technology Department is one of six departments within the School of Industrial and Engineering Technologies with the mission to provide students with the hands-on skills and knowledge required to succeed in the welding industry.

Currently the Welding Technology Department offers the following three programs of study:

- Welding (Certificate)
- Welding Technology (AAS)
- Welding & Fabrication Engineering Technology (BS)

All majors are supported by a common core of technical welding courses that encompass the most common welding and cutting processes in use by the welding industry.

Small classes enable our experienced faculty to give each student the personal attention necessary to optimize his or her learning experience. With over eighty years of combined industry and teaching experience, the welding faculty has a diversified background in the welding field.

The associate degree in welding technology curriculum offers practical skills and theory in welding, quality assurance, welding design, robotic welding, CNC plasma cutting, non-destructive testing (NDT), plus the mathematics, and language skills necessary to mature to a technical or management career in the welding and fabrication industries.

A graduate of this major should be able to do the following:

- Weld safely in shop and field operations
- Work safely and avoid practices that are unsafe to others
- Weld using oxy-fuel, electric, and inert gas shielded methods
- Distinguish the types of welding power sources, their characteristics, uses, and limitations
- Inspect welding jobs using visual, destructive, and non-destructive testing methods
- Construct weldments from sketches, blueprints, or verbal instructions
- Interpret welding symbols

- Select the proper welding process, procedure, supplies, etc., based on cost limitations
- Identify ferrous and nonferrous metals
- Test the physical and mechanical properties of metals, as related to weldability
- Follow welding qualification tests procedures according to specifications of the AWS, API, and ASME codes
- Program and operate CNC plasma oxy-fuel cutting equipment
- Program and operate a robotic weld station
- Apply the principles of metallurgy to the selection of a welding procedure

	Pennsylvania College of Technology	
Students Starting Fall 2011	Associate in Applied Science in Welding Technology	Total Credits: 67
Semester One		
WEL 113	Oxy-Fuel Welding and Cutting I	2
WEL 114	Shielded Metal Arc I	2
WEL 115	Oxy-Fuel Welding and Cutting II	2
WEL 116	Shielded Metal Arc II	2
WEL 102	Welding Blueprint and Layout	3
FYE 101	First Year Experience	1
SAF 110	Occupational Health and Safety	2
MTH 124 or MTH 180	Technical Algebra and Trigonometry I College Algebra and Trigonometry I	3
	Semester Total:	**17**
Semester Two		
WEL 120	Gas Metal Arc I	2
WE L123	Gas Tungsten Arc I	2
WEL 124	Gas Metal Arc II	2
WEL 129	Gas Tungsten Arc II	2

ENL 111	English Composition I	3
CSC 124	Information, Technology, and Society	3
FIT	Fitness and Lifetime Sports Elective	1
	Semester Total:	**15**
Semester Three		
WEL 210	Flux Cored and Sub-Arc I	2
WEL 230	Shielded Metal Arc III	2
WEL 214	Flux Cored and Sub-Arc II	2
WEL 234	Shielded Metal Arc V	2
WEL 240	Basic CNC Programming	3
ENL 201	Technical and Professional Communication	3
MSC 106	Introduction to Metallurgy	4
	Semester Total:	**18**
Semester Four		
WEL 213	Gas Tungsten Arc III	2
WEL 233	Shielded Metal Arc IV/Pipe Welding	2
WEL 219	Gas Tungsten Arc IV	2
WEL 239	Shielded Metal Arc VI/Pipe Welding	2
QAL 237	Non-Destructive Testing I	3
WEL 248	Robotic Welding	3
SSE	Social Science Elective	3
	Semester Total:	**17**
Welding Core Courses: WEL 113 - WEL 239 are two-credit, eight-week classes.		

The Bachelor of Science degree in welding and fabrication engineering technology is structured to support welding and joining operations where engineers pass plans and projects to mid-management personnel who must carry out the planning, organization, and delivery of manufacturing projects. Emphasis will be on developing skills needed to lead projects and interface with engineering and development teams. Students have the opportunity to prepare for careers in mid-management and supervisory positions, as well as technical positions, sales, service, or research.

A graduate of this major should be able to do the following:

- Solve manufacturing problems using computer hardware and software
- Solve manufacturing problems using scientific principles and methodology
- Demonstrate knowledge of safety and health in the occupation and in personal life
- Demonstrate knowledge of the impact and linkage of technology as a cultural universal
- Analyze and solve manufacturing problems of an economic, technical, organization, and design variety
- Analyze and recommend solutions of manufacturing problems of the moral, ethical, and legal nature
- Contribute to an industrial design team in the design, redesign, and upgrade of products to achieve improved manufacturability, aesthetics, and function

- Solve welding design and materials joining problems with accepted methods, processes, and techniques to meet industrial standards
- Demonstrate knowledge of legal and ethical behavior governing the engineering technologist
- Demonstrate skill in accepted welding and materials joining processes
- Interpret and use organizational economic and managerial techniques to promote profit, product, or service quality and institutional stability
- Demonstrate proficiency in welding automation, principles of fabrication, and process integration

	Pennsylvania College of Technology	
Students Starting Fall 2011	Bachelor of Science in Welding and Fabrication Engineering Technology	Total Credits: 66/67
Semester One		
FYE 101	First Year Experience	1
WEL 113	Oxy-Fuel Welding and Cutting I	2
WEL 114	Shielded Metal Arc I	2
WEL 115	Oxy-Fuel Welding and Cutting II	2
WEL 116	Shielded Metal Arc II	2
SAF 110	Occupational Health and Safety	2
WEL 102	Welding Blueprint and Layout	3
MTH 180	College Algebra and Trigonometry I	3
	Semester Total:	**17**
Semester Two		
WEL 120	Gas Metal Arc I	2
WEL 123	Gas Tungsten Arc I	2
WEL 124	Gas Metal Arc II	2
WEL 129	Gas Tungsten Arc II	2
CSC 124	Information, Technology, and Society	3
ENL 111	English Composition I	3

MTH 182	College Algebra and Trigonometry II	3
FIT	Fitness and Lifetime Sports Elective	1
	Semester Total:	**18**
Semester Three		
WEL 210	Flux Cored and Sub-Arc I	2
WEL 230	Gas Tungsten Arc III	2
WEL 214	Flux Cored and Sub-Arc II	2
WEL 234	Shielded Metal Arc V	2
WEL 240	Basic CNC Programming	3
ENL 201	Technical and Professional Communication	3
MTH 160 or MTH 230	Elementary Statistics with Computer Applications Applied Calculus	4/3
	Semester Total:	**17/18**
Semester Four		
WEL 213	Gas Tungsten Arc III	2
WEL 233	Shielded Metal Arc IV/Pipe Welding	2
WEL 219	Gas Tungsten Arc IV	2
WEL 239	Shielded Metal Arc VI/Pipe Welding	2
WEL 248	Robotic Welding	3
QAL 247	Non-Destructive Testing I	3
MSC 106	Introduction to Metallurgy	4
	Semester Total:	**18**
Semester Five		
SPC 101	Fundamentals of Speech	3
QAL 301	NDT Quality Assurance	3
CET 230	Statics	3
PHS 115	College Physics I	4
HIS 115 or HIS 125	World Civilization I World Civilization II	3
	Semester Total:	**16**

Semester Six		
WEL 301	Industrial Processes	3
CET 243	Strength of Materials	3
MET 311	Computer Solutions of Engineering Problems	3
MET 321	Engineering Ethics and Legal Issues	3
FIT	Fitness and Lifetime Sports Elective	1
OEA	Open Elective	3
	Semester Total:	**16**
Semester Seven		
WEL 410	Industrial Weld Design	3
WEL 400	Fabrication of Alloys	3
MET 495	Senior Seminar	1
EET 302	Industrial Electronics and Applications	3
HUM	Humanities Elective	3
SSE	Social Science Elective	3
	Semester Total:	**16**
Semester Eight		
WEL 420	Welding Codes and Procedures	3
MET 496	Senior Seminar - Lab	3
HUM, SSE, ART, AAE, IFE, or FOR	Humanities Elective Social Science Elective Art Elective Applied Arts Elective International Field Experience Elective Foreign Language Elective	3
ART	Art Elective	3
OEA	Open Elective	3
	Semester Total:	**15**
Welding Core Courses: WEL 113 & WEL 239 are two-credit, eight-week classes.		

Profile: *Michael Harris, Pennsylvania College of Technology*

Originally a mechanical engineering major, Penn College senior Michael Harris realized soon into his freshman year he was not enjoying that field and wanted a career that would not keep him confined to an office. A summer maintenance job introduced him to welding, and he switched his major to the welding engineering program.

"I love Penn College," says Harris. "In the classroom environment we have here, the instructors are more like friends. We are on a first-name basis, not so much a professor-student situation – you are on a personal level with them. It's easier for them to teach me like that." Harris also likes the fact the Penn College curriculum is heavier in application than theory. Regardless of degree path, all welding students get two years of manual welding classes. Harris says, "Most of us are certified pipe welders. It helps us out, being able to manually master our trade. The first year we take basic classes, the second is more advanced. The third and fourth years are engineering classes, statics, strengths and materials, ethics, advanced economics, and a quality assurance class."

Harris says the practical training has been his favorite part of the program, even though he came to Penn College with very little prior experience. In fact, his proudest moment so far has been passing his welder's certification exam (an optional exam for all welding students) on the first try with just two years classroom training and no prior experience. An enthusiastic Harris explains, "I was able to take everything I was taught, put it all together, and get something out of it right then and there." He also equates the time spent in his labs to going to a job four hours a day. "You were not confined to one classroom for four hours, but you were out doing projects. I liked that."

Conversely, he found his statics class to be one of the most difficult he had to take. "Statics is the study of forces and structural members. Loads, how many factors, where will it be dispersed evenly, things like that," says Harris.

The lucrative starting salaries for welding engineers, along with benefits like opportunity for overseas travel, makes swallowing statics class a little easier for Harris. Although he hasn't lined up a job yet, he has experienced a quality internship with American Car and Foundry, a manufacturer of railroad tank cars. During his internship, Harris shadowed X-ray and visual weld technicians and was also afforded the chance to work on some problems. For his future in this career, he has set his sights on the energy industry. Harris says, "I would like to be in the nuclear field, power generation, but I would also consider a refinery. I know I want to end up somewhere in the high-pressure industry. It interests me because so much is involved in fabricating for it. It takes years sometimes. It's very intricate, and you have to keep a close eye on every piece. I want to be in quality assurance. I am not a suit-and-tie individual. I like to be hands on. An X-ray or ultrasonic testing position would be ideal."

Bachelor of Science in Welding Engineering
The Ohio State University - Columbus, Ohio

The welding engineer is concerned with all of the activities related to the design, production, performance, and maintenance of welded products. Interest is primarily in the manufactured or fabricated product, including material selection, manufacturing methods, tooling, operator training, quality control (QC), performance evaluation, sales, and service. The broad range of welded products with which welding engineers deal includes structures, such as bridges and buildings; pressure vessels and heat exchangers, such as nuclear systems, boilers, chemical processing equipment, storage vessels and transmission and distribution piping; transportation vehicles for water, land, air, and space travel; and production and processing machines of all types. The welding engineering program provides basic liberal studies and the engineering training needed to function in the welding industry. Welding engineering courses at The Ohio State University (OSU) combine work in several engineering fields. The following four academic areas are addressed:

- **Design** – including work in engineering mechanics, stress analysis, structures, and machine and production design
- **Materials** – with coursework in physical metallurgy, metallography, and physical chemistry
- **Welding processes** – including electrical equipment and control
- **Fitness for service** – including non-destructive testing

Coursework in these four areas is taken in welding engineering and associated programs, the latter giving the student a perception of other engineering fields. Subsequent studies in the Department of Welding Engineering utilize this background information to provide in-depth training in welding materials, design, processes, and NDT evaluations. This is designed to prepare the student for complex research, production, and applications work in modern industry.

Seminars, field trips, and industrial experiences are included in the program.

The program's objectives are as follows:

- Welding engineers will be able to utilize the fundamental principles of engineering science and mathematics, and are aware of the underlying historic, social, ethical, and aesthetic aspects of engineering.
- Welding engineers will have knowledge of the fundamental theory of the process, design, materials, and testing aspects of welding.
- Welding engineers will be able to apply their fundamental welding engineering knowledge in an integrated fashion to solve diverse practical problems in the welding and joining field.
- Welding engineers will be able to communicate effectively in written, oral, and informal forms with a variety of audiences.
- Welding engineers will be able to work effectively in independent and collaborative aspects of their professional activity in an organized and productive fashion.

	The Ohio State University	
	Undergraduate Program in Welding Engineering	
Secondary Point Hour Courses:	You must have at least a 2.0 cumulative GPA and at least a 2.0 cumulative in these courses (OR their equivalent courses) to be admitted to the Welding Engineering major.	Credits:
MATH 151	Calculus and Analytic Geometry I	5
MATH 152	Calculus and Analytic Geometry II	5
MATH 153	Calculus and Analytic Geometry III	5
CHEM 121	General Chemistry	5
CHEM 125	Chemistry for Engineers	4
ENL 181	(or ENG H191) Data Acquisition	3
ENL 183	(or ENG H193) Engineering Fundamentals and Laboratory	3
EngGraph 167	(ENG H192, CSE 202 or CSE 294P)	4
Physics 131	Mechanics	5

Physics 132	Electricity and Magnetism	5
Other courses to be taken before beginning W.E. Major core courses:		
Physics 133	Waves	5
MATH 254	Calculus and Analytical Geometry IV	5
MATH 255 or 415	Differential Equations and Their Applications	5
	Ordering and Partial Differential Equations	4
MechEng 410	Statics	4
MechEng 420	Introduction Strength of Materials	4
ECE 300	Electrical Circuits	3
ECE 309	Electrical Circuits Lab	1
MSE 205	Introduction to Materials, Science, and Engineering	3
ISE 350	Manufacturing Engineering	3
ENGLISH 110	First-Year English Composition	4
WE 300	Survey of Welding Engineering	3
WE 350	Introduction Welding Lab 1	1
WE 351	Introduction Welding Lab 1	1

	The Ohio State University	
	Undergraduate Program in Welding Engineering	
Autumn:		Credits:
MSE 401	Materials Thermodynamics	4
WE 500	Physical Principles in Welding Engineering I	3
WE 550	Physical Principles in Welding Engineering I - Lab	1
WE 620	Engineering Analysis for Design and Simulation	4
WE 611	Welding Metallurgy I	3
WE 661	Welding Metallurgy I Lab	1
WE 690	Capstone Welding Design I	1
WE 489	Industrial Experience I	1
ISE 410	Industrial Quality Control	4
Tech Elec		
Winter:		

MSE 525	Phase Diagram	3
MSE 581	Materials Science Lab	2
WE 600	Physical Principles in Welding Engineering	3
WE 621	Welding Engineering Design	4
WE 612	Welding Metallurgy II	3
WE 662	Analysis of Non-Ferrous and High Alloy Welds	1
WE 691	Capstone Welding Design II	2
ISE 504	Engineering Economic Analysis	3
Tech Elec.		
Tech Elec.		
Spring:		
WE 601	Welding Process and Applications	3
WE 610	Introduction to Metallurgy	3
WE 631	Nondestructive Evaluation	4
WE 651	Welding Process Applications - Lab	1
MSE 543	Structural Transformations	3
WE 641	Welding Codes, Specifications, and Standards	3
WE 692	Capstone Welding Design II	2
Tech Elec.		
Tech Elec.		
Tech Elec.		
Tech Elec.		

The Welding Engineering degree program includes fourteen hours of Technical Electives. Electives must be approved by the Undergraduate Studies Committee prior to enrollment in the course(s). A correct, approved Technical Elective Approval Form must be on file before an Application to Graduate can be approved.

Your Technical Elective plan must show coherence and depth in Welding Engineering or in a closely aligned science or engineering discipline. A maximum of two credit hours of satisfactory/unsatisfactory graded courses will be approved for Technical Elective credit. The contribution of such courses to coherence or depth of the Technical Elective plan must be described in writing and accompany the Approval form.

	The Ohio State University	
	Recommended technical electives for Welding Engineering (Revised May 2009)	
Materials:		Credits:
WE 715	Weldability	3
WE 701	Solid State Welding	3
MSE 642	Polymer Science & Engineering	3
MSE 661	Ferrous Metallurgy	3
MSE 662	Corrosion	3
MSE 663	Non-Ferrous Metallurgy	3
Process:		
WE 605	Weld Process Control	3
WE 655	Weld Process Control Lab	1
WE 602	Fund of Resistance Weld Prc	3
WE 656	Robot Programming & Op	1
WE 701	Solid State Welding	3
WE 702	Fund of Resistance Weld	3
WE 703	Brazing & Soldering	3
WE 704	High Energy Density	3
WE 705	Adv Weld Proc Cont Sys	3
WE 706	Joining Polymer Comp	3
WE 707	Adhesive Bond & Mech Join	3
WE 755	Adv Weld Proc Cont Lab	1
ISE 601	Computer Apps in Indstr Ctrl	3
Design:		
WE 723	Analysis of Weld Syst	3
WE 740	Fitness for Service	3
ME 662	Intro Mech Composites	3

ME 762	Structural Composites	3
CE 431	Structural Eng. Principles	3
ISE 682	Design for Manufacturing	4
ISE 652	Rapid Prototyping Lab	3
NDE:		
WE 635	Fundamentals of Radiography	4
WE 638	Fundamentals of Ultrasonics	4
WE 681	NDE Seminar	1
WE 732	Ultrasonic NDE	4
ME 682	Exp. Methods in Mechanics	3
Polymers:		
WE 706	Welding Plastics and Comp	3
WE 707	Adhesive Bondg & Mech Jng	3
MSE 642	Polymer Sci & Engineering	3
ISE 751.01	Intro to Modlng Mtrls Proc	3
ISE 751.02	Polymer Proc Fundmtls	3
ISE 652.01	Rapid Prototyping Lab	3
ME 662	Intro to Mech of Comps Stat	3
ME 762	Structural Composites	3
ChemE 773	Intro Hi Polymer Eng	3
ChemE 776	Princ Polymer Conversn	3
Other:		
WE 736	Health & Safety	3
WE 795	Graduate Seminar	1
ISE 681	Project Management	3
NucEng 505	Intro to Nuclear Eng	3
NucEng 606	Radiological Safety	3
WE H783	Seniors Honors Research	
WE 694	(1-5 as offered) Group Studies	
WE 695	(1-3 as offered) WE Topics	
WE 793	Individual Study	

Profile: *Melissa Rubal, OSU*

OSU graduate student Melissa Rubal went through high school knowing she wanted to be an engineer, but wasn't sure exactly which kind. Along the way she took every pre-engineering and AP level class she could and did her research on the different engineering fields available. In the end, she chose welding engineering because it encompassed a variety of engineering disciplines. Rubal says, "What I really like about this field is that we have to take so many other engineering classes, so I looked at it as a spectrum introducing me to a variety of subjects and skills."

She says when she tells people her career field, she often gets "funny looks," but so far has not encountered any discrimination in her studies or at her internship, which she completed at Lincoln Electric Company in Cleveland, Ohio. She attributes the strange looks to the fact that people often misunderstand welding engineering to be the same as welding. "I only had to take two welding classes," says Rubal, a hint of relief in her voice. "Welders have to learn the skills behind the job, but we focus on how it all works. We know the breakdowns of the machines, the physics behind the welding, heat flow, material selection, and joint design. We are the problem solvers and designers, and you need actual welders to perform that function. Even though we take our welding classes, most of us aren't good usually, so we definitely respect the manual skills welders."

While welding lab did not rank as one of Rubal's favorite classes, metallurgy did. In fact, as part of her graduate studies program, which Rubal began dually in her senior year, she works in the metallurgy lab. "I like learning about the processes and seeing how things look under the microscope," she says. In fact, in October 2008, Rubal presented her first international conference paper in Pittsburgh, Pennsylvania. Her chosen topic was friction stir processing of titanium.

Well on her way to an exciting career, Rubal recommends the welding engineering program at OSU to students who already have an engineering and math/science aptitude. "The instructors here obviously really know what they're talking about," she says. "The teachers and graduates of this program have international recognition. It's a big school, but a small program. You will not get lost."

Rubal believes students who prepare for an engineering field while still in high school will have an advantage in this program. Rubal says, "I recommend developing good study habits and being a hard worker. You don't have to have welding experience when you come here, but you do need study skills, and you need to be prepared to work hard. Take the most advanced classes offered while you're in community college or high school."

No stranger to hard work herself, Rubal is exploring several career paths at this time, including possibly Research and Development or teaching. "I'm keeping my options open right now," she says. "I know I want to be a problem solver. I don't want to sit at a desk and do the same thing every day."

Online Welding Resources

General Information

Welding and Metals Joining Resources

metals.about.com/od/weldin1/Welding_and_Metals_Joining_Resources.htm

Welding Terms

www.millerwelds.com/resources/dictionary.html

Welding, Cutting, and Brazing

www.osha.gov/SLTC/weldingcuttingbrazing

Organizations

American Welding Society

www.aws.org

Weld-Ed

www.weld-ed.org

The Welding Institute

www.twi.co.uk

Welding Employers

Acute Technological Services

www.acutetechserv.com

Bechtel Corporation

www.bechtel.com

Fluor Corporation

www.fluor.com

Zachry Construction

www.zachryconstructioncorp.com

Welding Blogs

Miller Industry News Blog

www.millerwelds.com/results/blog

This blog covers topics involving the welding trade and ranging from current events to new product announcements. Features include how-to videos and the option for members to submit topic ideas.

WCWelding Welding Tips and Projects Blog

www.wcwelding.com/welding-tips-blog.html

Especially helpful for entry-level welders, this site offers how-to's for improving welding skills. The site provides information on welding careers, industry news, and other metalworking opportunities.

Welding Advisors Update Blog

www.welding-advisers.com/Welding-plan-blog.html

Subjects featured on this blog include welding processes and equipment, jobs, and safety regulations. The site offers links to browse new equipment for online purchase.

Stick Welding, MIG Welding, and TIG Welding Blog

www.weldingtipsandtricks.com/Welding-Tips-blog.html

This blog offers information on stick welding, metal inert gas (MIG) welding, and tungsten inert gas (TIG) welding. Readers can submit questions to be answered in daily blog posts. Videos and pictures demonstrate welding techniques and mechanical processes. Members also can join the welding discussion forum.

***Welding & Gases Today* Editor's Blog**

www.weldingandgasestoday.org/blog/

Developed by the Gases and Welding Distributors Association, this site provides news updates affecting the welding industry. Subjects include gas and oil news, new products, and management tips. The blog is a good source for all levels of welding experience.

Welder World Blog

welderworld.com/blog/

This site features news updates and question forums for welding professionals. Blog posts cover assessments and explanations of new products, tools, and machinery. Also featured are how-to videos, books, and articles. Entertainment materials, such as movies and books that feature welding themes, also are included.

Paul Horton's Welder Series Blog

www.welderseries.com/blog/

Developed by welding professional Paul Horton, this blog details Welding Series products. Posts cover consumer reviews, how-to's, video, and pictures for easy learning. Welding news and events also are covered. Shopping links are provided for purchasing equipment.

Welding Design & Fabrication News

weldingdesign.com/news/

Articles provided on this site detail welding business and news. Reports on manufacturing, commerce, and industry developments are supplied for professional reading. The site includes links to welding newsletters and online shopping guides.

WeldingWeb Forum

weldingweb.com/index.php

An interactive forum for new and experienced welders, this site gives experts the chance to ask and answer questions. Members can share ideas, tips, and advice on welding processes and tools.

Conclusion

In today's economy, there is much anxiety about the stability and future of any job and its related career pathways. However, for those with the interest, skills, and desire, welding can offer long-term and lucrative rewards. Whether it's building the next generation of nuclear reactors, skyscrapers, or undersea oil rigs, welders are the lifeblood of construction all over the world. By finding and completing the right training, you too can take advantage of the many opportunities this field has to offer.

Index

A

B

C

E

F

G-H

J

L-M

N

O

P

R-S

T

W

Z

About the Authors

Joseph Abbott

Joseph Abbott is a writer and copy editor based in Waco, Texas. A University of Texas graduate, he started his newspaper career at the UT student newspaper, *The Daily Texan*, and went on to the copy desk at the *Waco Tribune-Herald*. He is a spaceflight enthusiast who has seen (and covered for wacotrib.com) two shuttle launches in person.

Karen Mitchell Smith

Karen Mitchell Smith earned her English degree from Texas Tech University in 1985 and, since then, has been a college recruiter, teacher and longtime writer. An accomplished motivational speaker, she also presents seminars for educators and high school students.

Established in 2004, TSTC Publishing is a provider of high-end technical instructional materials and related information to institutions of higher education and private industry. "High end" refers simultaneously to the information delivered, the various delivery formats of that information, and the marketing of materials produced. More information about the products and services offered by TSTC Publishing may be found at its website: publishing.tstc.edu.

TechCareers Series

TSTC Publishing launched the *TechCareers* series with *Biomedical Equipment Technicians* in 2008. TSTC Emerging Technologies initially underwrote the series, created to inform the public about existing technologies and those to come.

TSTC Emerging Technologies also provided funding for 500 copies of each book in the series to be distributed throughout Texas to high school career and technical education counselors. In addition to *Biomedical Equipment Technicians,* the series includes *Automotive Technicians, Wind Energy, Avionics* and *Computer Gaming Programmers and Artists*. Forthcoming titles include: *Aviation Maintenance, Aviation Pilots*, and *Graphic Design*. For information about TSTC Emerging Technologies, go to forecasting.tstc.edu.

Every TechCareers *book features:*

- Detailed overviews of career pathways, skill sets, and educational requirements
- Profiles of professionals, experts, employers, current students, and instructors
- Program listings, sample degree plans, and additional industry resources
- Salary ranges and benefits

TechCareers: Wind Energy

By Mike Jones

$14.95 Softback
ISBN 978-1-934302-55-2
$9.99 Ebook
ISBN 978-1-936603-02-2
1st edition
August 2010

Kindle edition available at Amazon.com

"With increasing demand for wind power, there is already a pressing need for wind technicians to maintain the wind turbines."

-TechCareers: Wind Energy

TechCareers: Wind Energy describes the current wind energy market and the explosive growth of the energy field. Due to current interest in green energy, wind is now becoming a popular energy source in countries across the world. Increased demand for expert wind energy professionals is expected to continue. This book describes the jobs needed to support this growing career field and the education and necessary skills for success.

TechCareers: Avionics

By Helen Ginger

$14.95 Softback
ISBN 978-1-934302-47-7
1st edition
October 2009

"If you're just beginning to think about this field and a career in avionics, now is the opportune time to begin your training ... the good news for those now considering this field is that the super techs are reaching retirement age, creating a job gap and a wide open door for new techs to step through."

-TechCareers: Avionics

TechCareers: Avionics gives a clear description of the growing need for avionics technicians. Avionics deals with the maintenance and repair of all flight instruments, including flight control, weather radar, and missile control. With many of the current avionics technicians reaching retirement, airlines will be looking for new techs to take their places. Featuring necessary information about job opportunities, training, and educational requirements, *TechCareers: Avionics* is a valuable tool for prospective students entering the field.